T0212331

SpringerBriefs in Computer Science

Series editors

Stan Zdonik, Brown University, Providence, RI, USA
Shashi Shekhar, University of Minnesota, Minneapolis, MN, USA
Xindong Wu, University of Vermont, Burlington, VT, USA
Lakhmi C. Jain, University of South Australia, Adelaide, SA, Australia
David Padua, University of Illinois Urbana-Champaign, Urbana, IL, USA
Xuemin Sherman Shen, University of Waterloo, Waterloo, ON, Canada
Borko Furht, Florida Atlantic University, Boca Raton, FL, USA
V. S. Subrahmanian, Department of Computer Science, University of Maryland,
College Park, MD, USA
Martial Hebert, Carnegie Mellon University, Pittsburgh, PA, USA
Katsushi Ikeuchi, University of Tokyo, Tokyo, Japan
Bruno Siciliano, Università di Napoli, Naples, Italy
Sushil Jajodia, George Mason University, Fairfax, VA, USA
Newton Lee, Newton Lee Laboratories, LLC, Tujunga, CA, USA

SpringerBriefs present concise summaries of cutting-edge research and practical applications across a wide spectrum of fields. Featuring compact volumes of 50 to 125 pages, the series covers a range of content from professional to academic.

Typical topics might include:

- A timely report of state-of-the art analytical techniques
- A bridge between new research results, as published in journal articles, and a contextual literature review
- A snapshot of a hot or emerging topic
- An in-depth case study or clinical example
- A presentation of core concepts that students must understand in order to make independent contributions

Briefs allow authors to present their ideas and readers to absorb them with minimal time investment. Briefs will be published as part of Springer's eBook collection, with millions of users worldwide. In addition, Briefs will be available for individual print and electronic purchase. Briefs are characterized by fast, global electronic dissemination, standard publishing contracts, easy-to-use manuscript preparation and formatting guidelines, and expedited production schedules. We aim for publication 8–12 weeks after acceptance. Both solicited and unsolicited manuscripts are considered for publication in this series.

More information about this series at http://www.springer.com/series/10028

Lili Mou · Zhi Jin

Tree-Based Convolutional Neural Networks

Principles and Applications

Springer

Lili Mou
AdeptMind Research
Toronto, ON, Canada

Zhi Jin
Institute of Software
Peking University
Beijing, China

ISSN 2191-5768   ISSN 2191-5776 (electronic)
SpringerBriefs in Computer Science
ISBN 978-981-13-1869-6   ISBN 978-981-13-1870-2 (eBook)
https://doi.org/10.1007/978-981-13-1870-2

Library of Congress Control Number: 2018953306

© The Author(s) 2018
This work is subject to copyright. All rights are reserved by the Publisher, whether the whole or part of the material is concerned, specifically the rights of translation, reprinting, reuse of illustrations, recitation, broadcasting, reproduction on microfilms or in any other physical way, and transmission or information storage and retrieval, electronic adaptation, computer software, or by similar or dissimilar methodology now known or hereafter developed.
The use of general descriptive names, registered names, trademarks, service marks, etc. in this publication does not imply, even in the absence of a specific statement, that such names are exempt from the relevant protective laws and regulations and therefore free for general use.
The publisher, the authors and the editors are safe to assume that the advice and information in this book are believed to be true and accurate at the date of publication. Neither the publisher nor the authors or the editors give a warranty, express or implied, with respect to the material contained herein or for any errors or omissions that may have been made. The publisher remains neutral with regard to jurisdictional claims in published maps and institutional affiliations.

This Springer imprint is published by the registered company Springer Nature Singapore Pte Ltd.
The registered company address is: 152 Beach Road, #21-01/04 Gateway East, Singapore 189721, Singapore

To all

Preface

In recent years, neural networks have become one of the most popular models in various applications of artificial intelligence, including image recognition, speech processing, and natural language processing. The convolutional neural network and the recurrent neural network are among the most popular neural architectures. The former uses a sliding window to capture translation invariant features; it typically works with signals in a certain dimensional space (e.g., 1D speech or 2D image). The latter is suitable to process time-series data as it iteratively aggregates information.

However, such models cannot explicitly incorporate more complicated structures, e.g., the parse tree of a sentence. Socher et al. propose a recursive neural network that propagates information recursively bottom up along a tree structure. Although the recursive network can encode the tree structure to some extent, it has a long propagation path and may suffer from the problem of gradient exploding or vanishing during training.

In this book, we propose a new neural architecture, the *tree-based convolutional neural network* (TBCNN). It combines the merits of convolutional neural networks and recursive neural networks. Our key idea is to design a subtree feature detector, applied to different regions of a tree. Then, a dynamic pooling layer aggregates information to a fixed-size vector for further processing. In this way, TBCNN has short propagation path as a convolutional neural network, but is structure sensitive as a recursive neural network.

We would like to take you a tour to three applications of TBCNN: the abstract syntax trees of program source code, the constituency trees of natural language sentences, and the dependency trees of natural language sentences. In each application, we will first introduce the background of the domain, and then design a TBCNN variant especially suited to the trees in that domain. In the design, we will particularly address three technical difficulties, namely, the representations of nodes, the weight assignment of convolution, and the way of pooling. These provide useful philosophy of designing TBCNN variants as well as other neural models. We will also present detailed experiments, showing that TBCNN has achieved high performance in all these domains.

Our book addresses both rigorous mathematics and intuitive understanding. The book is suitable for researchers and graduate students in the fields of machine learning and natural language processing. It is also suitable for general public who are interested in machine learning, or more generally, artificial intelligence. The foundations of generic neural networks are covered in Chap. 2, and for each application domain, we have made efforts to make it self-contained by introducing sufficient background knowledge. We hope this book would be an interesting read for a wide range of audience.

Toronto, Canada   Lili Mou
Beijing, China   Zhi Jin
July 2018

Acknowledgements

This research is supported by the National Basic Research Program of China (the 973 Program) under Grant No. 2015CB352201, and the National Natural Science Foundation of China under Grant Nos. 61620106007 and 61751210. The research in the book started in early 2014, and was conducted mainly at the Key Laboratory of High Confidence Software Technologies (Peking University), Ministry of Education; and the Institute of Software, Peking University. During the writing of the book, the first author had a post-doctoral fellow position at the University of Waterloo, and then took a scientist position at AdeptMind Research. We would like to thank the University of Waterloo and AdeptMind Inc. (Toronto) for their support of the writing of the book as well as research in general.

The authors would like to thank Prof. Lu Zhang, Dr. Ge Li, Dr. Yan Xu, Dr. Yangyang Lu, Hao Peng, Yuxuan Liu, Hao Jia, Ran Jia, Rui Men, and Zhao Meng at Peking University for their help in the research. We would also like to thank Dr. Lixing Li, Dr. Yunchuan Chen, and Dr. Weizhuo Li from the University of Chinese Academy of Sciences; together with them and Dr. Yan Xu and Dr. Yangyang Lu, we self-organized weekly seminars, during which we have learned both foundations and frontiers of machine learning research.

In our research, we have also discussed with many other colleagues, including Dr. Hang Li from Toutiao AI Lab, Dr. Zhengdong Lu from DeeplyCurious.ai, and Yiping Song, Bolin Wei, Jingsi Wen, Wenhao Huang, and Zhenxin Fu from Peking University. We would like to thank them for their insightful thoughts.

Last but not least, we would like to thank our families and friends for their long-time help and support.

Thank you all!

Contents

Acronyms

AI	Artificial Intelligence
ANN	Artificial Neural Network
API	Application Programming Interface
AST	Abstract Syntax Tree
BoW	Bag-of-Words
CBoW	Continuous Bag-of-Words
CCG	Combinatory Categorical Grammar
CNN	Convolutional Neural Network
CRF	Conditional Random Field
DAG	Directed Acyclic Graph
DNN	Deep Neural Network
GRU	Gated Recurrent Unit
LSTM	Long Short-Term Memory
NLP	Natural Language Processing
MLP	Multilayer Perceptron
MSE	Mean Square Error
OJ	Open Judge
PCA	Principle Component Analysis
RBM	Restricted Boltzmann Machine
ReLU	Rectified Linear Unit
RGB	Red, Green, and Blue
RBF	Radius-Basis Function
RNN	Recurrent Neural Network
SGD	Stochastic Gradient Descent
SVM	Support Vector Machine
TBCNN	Tree-Based Convolutional Neural Network
c-TBCNN	Constituency Tree-Based Convolutional Neural Network
d-TBCNN	Dependency Tree-Based Convolutional Neural Network

Chapter 1
Introduction

Abstract In this chapter, we provide a whirlwind introduction of the history of deep neural networks (also known as *deep learning*), positioned in a broader scope of machine learning and artificial intelligence. We then focus on a specific research direction of deep neural networks—incorporating structural information of data into the design of network architectures. This motivates the key contribution of the book, a tree-based convolutional neural network (TBCNN), that performs the convolution operation over tree structures. Finally, we provide an overview of this book.

Keywords Deep learning · Tree-based convolution · Structure modeling

1.1 Deep Learning Background

Artificial Intelligence (AI) has long been a hot research topic in computer science. According to Nilsson, AI is "concerned with intelligent behavior in artifacts," and "intelligent behavior, in turn, involves perception, reasoning, learning, communicating, and acting in complex environments" [7]. The mission of AI research can be divided into two schools: strong AI and weak AI. The former seeks to obtain true intelligence like humans, whereas the latter focuses on developing intelligent algorithms for a particular application. Penrose is a strong opponent of strong AI, and his book, *The Emperor's New Mind*, provides an interesting discussion on this topic [8]. There could be different arguments for the nature of intelligence, among which are *symbolicism*, *connectionism*, and *behaviorism*. Although it is not the core of this book, we would like to point out that *connectionism* mainly refers to artificial neural networks (ANN); it believes that intelligence lies in the connection of neurons.

In terms of AI applications, *machine learning* is the major technique for most modern intelligent systems, and has wide applications in computer vision, speech processing, and natural language processing (NLP). Several common (and perhaps simplified) machine learning paradigms include the following:

- *Supervised learning*. In a particular data point, one variable is of particular interest, and we denote it by y; other variables are denoted as x. If x is a vector or (a scalar), it is also known as *feature(s)*. Supervised learning is provided with a training set

© The Author(s) 2018
L. Mou and Z. Jin, *Tree-Based Convolutional Neural Networks*,
SpringerBriefs in Computer Science, https://doi.org/10.1007/978-981-13-1870-2_1

comprising known pairs of x and y. Given a new data point x_*, the goal is to predict its label y_*. If y is a continuous variable, we call the supervised learning *regression*; if y is categorical, we call it *classification*. An example of supervised learning is hand-written digit recognition: given a 10×10 image (100 pixels in total), we would like to predict which digit among 0 to 9 it is. In this sample, $x \in \mathbb{R}^{100}$ and $y \in \{0, 1, \ldots, 9\}$. Since y is a categorical variable here, this is a classification task.

- *Unsupervised learning.* In this paradigm, a data point does not have a variable (or label) of particular interest. The goal of unsupervised learning is to capture internal properties of data. Common examples of unsupervised learning include clustering, anomaly detection, and dimensionality reduction. In the regime of deep learning, another example of unsupervised learning is vector representation learning. In NLP tasks, neural networks typically work with word vectors (also known as *word embeddings*), which map discrete words to real-valued vectors. These embeddings are usually learned from unlabeled corpora, e.g., Wikipedia,[1] and hence word embedding learning is an unsupervised algorithm.

- *Reinforcement learning.* Similar to supervised learning, reinforcement learning (RL) aims to predict the labels based on data x, although in most cases RL models sequential prediction, i.e., a sequence of labels $y = y_1, y_2, \ldots, y_t$ over time steps. What differs from supervised learning is that, in RL, the true labels $y_1, \ldots, y_t$ are unknown, but some reward will be given based on a prediction sequence. By maximizing the reward, the system can have better predictions of y. Readers interested in RL, especially deep learning-based RL, are referred to [6].

Among the large number of machine learning models is the artificial neural network (ANN). In particular, ANN is biologically inspired: in 1958, Rosenblatt proposed the *perceptron* model [11], which mimics the dendrites activating an axon; a multilayer neural network can then be viewed as a stack of perceptrons, analogous to the connections of neurons in human brains.

Although biologically inspired, neural networks did not attract much attention at the beginning, largely due to Minsky and Papert's book, *Perceptrons* [5]. In the book, the authors provided detailed analysis to Rosenblatt's perceptrons, and proved some of perceptrons' drawbacks, for example, linearity. That is to say, perceptrons could only solve linearly separable classification problems, which is too strict in most real applications. The authors suspected linearity also holds for multilayer neural networks, which aroused concern in the community on the capability of neural networks. Interested readers are referred to [1, Sect. 1.7] for the history.

In 1980s, backpropagation was proposed to analytically compute the gradient of weights in a multilayer perceptron, making it possible to train a two-layer neural network efficiently [10]. This also proves Minsky and Papert's conjecture incorrect. However, due to a problem known as *gradient vanishing or exploding*, backpropagation did not work well with deep networks, whose application was still limited at that time.

Moreover, the support vector machine (SVM) with nonlinear kernels were proposed in 1990s, and its performance was significantly better than a neural network at

---

[1]https://www.wikipedia.org/.

that time. As a result, neural networks were not thought of as a serious machine learning model until the early of this century, which in turn suppresses the development of neural networks.

In 2006, Hinton et al. [2] proposed a pretraining method for neural networks, which *pretrains* neural weights with a restricted Boltzmann machine (RBM) in a layer-wise fashion. This makes the training of neural networks both efficiently and effectively. On the other hand, Internet technology (e.g., crowdsourcing) makes it possible to obtain massive labeled data, and as some experiments indicate, should the dataset be large enough, deep neural networks would achieve high performance even without pretraining techniques. These facts largely advanced the research of deep neural networks (also known as *deep learning*). Image recognition and speech processing are the first two domains where deep learning made significant breakthroughs; recently, deep learning has also been widely adopted in NLP and achieved high performance in various tasks.

1.2 Incorporating Structural Information into Neural Architectures

With the wide applications of deep learning, one research direction is to incorporate structural prior into the design of neural architectures, for example, 2D images [4], text with parsing trees [12], and graph-structured molecules https://pubs.acs.org/doi/full/10.1021/acs.jcim.5b00238 and social networks [9]. These models are analyzed in detail in Chap. 2, but we will get a glimpse at the general idea in this section.

In the field of image processing, for example, the image is typically a two-dimensional (2D) matrix, each element of which represents the intensity of a pixel. Since basic semantic units—e.g., a car, a tire, or even a stripe in the tire—are manifested with a set of neighboring pixels, people design a sliding window to capture local features. These local features are slided over different part of an image with weights being fixed, so that they have the *translation invariance* property. Such architecture is known as a *convolutional neural network* (CNN). In NLP, CNN can also be applied to natural language sentences, as its convolution window captures the semantic of neighboring words.

Recurrent neural networks (RNNs), different from CNNs, are capable of capturing sequential data. For example, a sentence can be viewed as some signal in the time series. An RNN reads a word at a time and maintains its state according to its previous state as well as the current input.

Socher et al. [12] proposed a recursive neural network[2] to explicitly model the internal structure of data, e.g., a constituency tree or a dependency tree of a natural language sentence. The idea is to represent each node as a vector recursively in a

---

[2]In the literature, a recursive neural network is sometimes also abbreviated as RNN. However, this is confusingly the same as a recurrent neural network. We do not use this acronym for the recursive neural network.

bottom-up fashion from leave nodes to the root. Then, the root node's vector is some representation of the entire tree, which can be used for the task of interest, e.g., sentiment classification.

Although a fully connected multilayer neural network will become more and more powerful if we add more neurons or layers, such architecture does not capture the internal structure of data in an explicit fashion. By contrast, those structure-sensitive neural networks leverage prior knowledge of data structures, serving as an induced bias that makes the learning machine more suited to the task.

1.3 The Proposed Tree-Based Convolutional Neural Networks

Despite the wide applications of neural networks with convolutional, recurrent, and recursive structures, they have their own drawbacks.

A CNN or RNN cannot make use of more complicated structures of data than spatial or temporal neighboring, e.g., the parse tree of a sentence. Although some evidence shows RNNs are more or less capable of learning syntax by itself [3], it also implies that the RNN has to learn a parser implicitly, and in some cases especially when the dataset is small, this could be a disadvantage.

A recursive neural network explicitly uses the tree structure of a data sample as the network's skeleton. However, if the tree is large, the recursive net has long propagation paths, which further cause the problem of gradient exploding or vanishing. Another potential problem of the recursive net is that, in some applications, leaf nodes are more important than intermediate nodes (e.g., the constituency parse tree of a sentence). The recursive structure buries illuminating leaf nodes at the bottom of its structure, and along propagation, such information may blur or get distorted.

Table 1.1 summarizes the pros and cons of existing models. The research question of this book is—Can we combine the merits of both? On the one hand, we would like to have short propagation path as convolutional neural networks, which allows efficient learning and effective information propagation. On the other hand, we would like to explicitly make use of tree structures of data as recursive neural networks.

To this end, we propose a *tree-based convolutional neural network* (TBCNN) in this book. The idea is to design a subtree feature detectors, sliding over an entire tree to

Table 1.1 Comparison of convolutional and recursive neural networks

Model	Advantage	Disadvantage
Convolutional	Short propagation path Efficient learning	Cannot make use of tree structural information
Recursive	Make use of prior knowledge of tree structures	Long propagation path Less efficient learning

extract local structural information. Then, dynamical pooling techniques aggregate extracted information from different parts of the tree. Such information serves as the vector representation of the tree and can be used in the task of interest. With this design philosophy, TBCNN would have short propagation path as conventional CNNs, while it is capable of effectively extracting structural features.

In this book, we evaluate the effectiveness of TBCNN in two domains: program analysis and natural language processing (NLP). In particular, we apply TBCNN to the abstract syntax tree of a program to classify programs' functionality and to detect certain patterns of source code. In NLP, we apply it to two major parsing trees (constituency trees and dependency trees) of a sentence with tasks of sentiment analysis, question classification, and natural language inference. Based on the properties of tree structures, we would design TBCNN variants, so that it can be adopted to different domains and achieves high performance in various tasks.

To sum up, the main contributions of this book include the following:

- We propose the novel tree-based convolutional neural network (TBCNN). With sliding windows of subtrees, TBCNN is able to capture structural features as a recursive neural network; with short propagation paths as in conventional CNNs, such features can be effectively learned by backpropagation.
- We apply TBCNNs to three types of trees: the abstract syntax tree of a program, the constituency tree of a natural language sentence, and the dependency tree of a sentence. By designing model variants of TBCNN, we achieve higher performance than previous state-of-the-art models in program classification, sentiment analysis, and question classification. In a natural language inference task, TBCNN outperforms previous sentence encoding-based approaches, and achieves comparable performance to word-by-word attention models, whose matching complexity is higher than TBCNN in order.
- To the best of our knowledge, we are also the first to apply modern neural networks to program analysis, with applications of classifying programs' functionality and detecting source code patterns. In this book, we would explore both unsupervised vector representation methods and supervised classification tasks in the field of program analysis.

1.4 Structure of the Book

Figure 1.1 shows the mindmap of this book.

As an introduction, this chapter introduces the origin and recent progress of neural networks. We also address the research topic of incorporating prior knowledge of data structures into the design of neural architectures. Then, we describe the proposed tree-based convolutional neural network and highlight the contributions of our work, concluded by an overview of the book.

Chapter 2 presents background knowledge and reviews related work in the literature. First, we introduce the traditional multilayer neural network and its learning

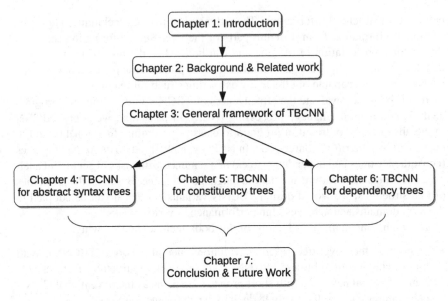

Fig. 1.1 Mindmap of this book

algorithms. Then, we will have a closer look at the neural networks in the field of NLP. Since this book mainly deals with programming language and natural language, these models will be more relevant to the topics in this book. Finally, we introduce, in detail, several structure-sensitive neural architectures, namely, convolutional, recurrent, and recursive neural networks, which will motivate our TBCNN model.

Chapter 3 proposes the general framework of TBCNN. We present the generic formula of tree-based convolution and discuss the difficulties of designing TBCNN in different scenarios. Then, we apply TBCNN to several fields: abstract syntax trees of programs, the constituency, and dependency trees of natural language sentences, which will be discussed in Chaps. 4, 5, and 6, respectively. In each chapter, we introduce the application background, describe the TBCNN variant, and present experimental results.

Chapter 7 concludes the book and discusses future work.

References

1. Haykin, S.S., Haykin, S.S., Haykin, S.S., Haykin, S.S.: Neural Networks and Learning Machines. Pearson Education (2009)
2. Hinton, G., Osindero, S., Teh, Y.: A fast learning algorithm for deep belief nets. Neural Comput. **18**(7), 1527–1554 (2006)
3. Karpathy, A., Johnson, J., Li, F.F.: Visualizing and understanding recurrent networks (2015). arXiv preprint arXiv:1506.02078

4. Krizhevsky, A., Sutskever, I., Hinton, G.: ImageNet classification with deep convolutional neural networks. In: Advances in Neural Information Processing Systems, pp. 1097–1105 (2012)
5. Minsky, M., Papert, S.: Perceptrons. MIT Press (1969)
6. Mnih, V., Kavukcuoglu, K., Silver, D., et al.: Human-level control through deep reinforcement learning. Nature **518**(7540), 529 (2015)
7. Nilsson, N.J.: Artificial Intelligence: A New Synthesis. Elsevier (1998)
8. Penrose, R.: The Emperor's New Nind: Concerning Computers, Minds, and the Laws of Physics. Oxford Paperbacks (1999)
9. Perozzi, B., Al-Rfou, R., Skiena, S.: DeepWalk: Online learning of social representations. In: Proceedings of the 20th ACM SIGKDD International Conference on Knowledge Discovery and Data Mining, pp. 701–710 (2014)
10. Remulhar, D., Hinton, G., Williams, R.: Learning representations by back-propagating errors. Nature **323**(9), 323–533 (1986)
11. Rosenblatt, F.: The perceptron: A probabilistic model for information storage and organization in the brain. Psychol. Rev. 832–837 (1958)
12. Socher, R., Pennington, J., Huang, E., Ng, A., Manning, C.: Semi-supervised recursive autoencoders for predicting sentiment distributions. In: Proceedings of the Conference on Empirical Methods in Natural Language Processing, pp. 151–161 (2011)

Chapter 2
Background and Related Work

Abstract In this chapter, we introduce the background of neural networks and review related literature. Section 2.1 introduces the general neural network and its learning algorithm—backpropagation. Section 2.2 addresses specialty of natural language processing, and introduces neural language models and word embedding learning. Section 2.3 introduces existing structure-sensitive neural networks, including the convolutional neural network, recurrent neural network, and recursive neural network.

Keywords Neural network · Neural language modeling · Word embeddings
Convolutional neural network · Recurrent neural network
Recursive neural network

2.1 Generic Neural Networks

2.1.1 Neuron and Multilayer Network

A *neuron*, also known as a *perceptron*,[1] a *logit*, or a *unit*, is the basic component of a neural network.

Let the input feature be $x \in \mathbb{R}^n$, where n is the number of features. The output of a neuron is computed by the following two steps:

- Computing the weighted sum of each dimension of x, and the weights $w \in \mathbb{R}^n$ are the parameters of the neuron. A biased term $b \in \mathbb{R}^n$ can be added to the weighted sum, but this is not the key of the neuron.
- An *activation function* is applied to the above weighted sum.

This process is illustrated in Fig. 2.1. Generally, the output of a neuron, denoted by y, is given by

---

[1] The orthodox perceptron, introduced by Rosenblatt [40], only uses thresholding as the activation function, that is, if the weighted sum of input is less than a threshold, the perceptron outputs 0, or otherwise, 1. In this sense, the perceptron is a special type of neuron. However, we do not distinguish these two terminologies as they are very similar.

© The Author(s) 2018
L. Mou and Z. Jin, *Tree-Based Convolutional Neural Networks*,
SpringerBriefs in Computer Science, https://doi.org/10.1007/978-981-13-1870-2_2

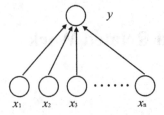

Fig. 2.1 A neuron. In the figure, z is omitted, which is a common simplified representation of a neuron

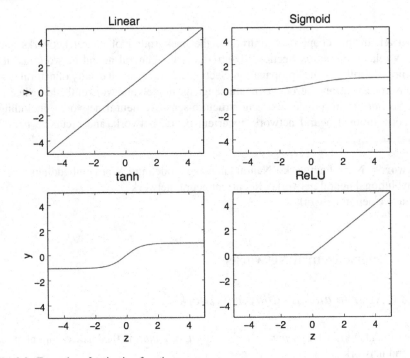

Fig. 2.2 Examples of activation functions

$$z = \mathbf{w}^\top \mathbf{x} + b \qquad (2.1)$$
$$y = f(z) \qquad (2.2)$$

Figure 2.2 plots several common activation functions, including:

- Linear function.

$$f(z) = z \qquad (2.3)$$

For linear activation, the slope is not needed (like $f(z) = k \cdot z$) because it can be absorbed in the weights in Eq. (2.1). That is to say, the activation function can tolerate a scaling factor (constant) for input.

- Sigmoid function (also known as logistic function).

$$f(z) = \sigma(z) \overset{\text{def}}{=} \frac{1}{1 + e^{-z}} \tag{2.4}$$

For binary classification, sigmoid is typically used as the activation of the output. This is derived by an assumption of the generalized linear model of x and y. If a sigmoid neuron is directly applied to input features, we will obtain the logistic regression model. For multi-class classification, a softmax function can be applied as the activation function. For the details, readers can refer to any machine learning textbook (e.g., [4]).

- Hyperbolic tangent.

$$f(z) = \tanh(z) = \frac{e^z - e^{-z}}{e^z + e^{-z}} \tag{2.5}$$

Hyperbolic tangent is a scaling of the sigmoid function in terms of both input and output. However, it may be easier to train because its output is in the range $(-1, 1)$, balancing positive and negative values.

- Rectified linear unit (ReLU).

$$f(z) = \begin{cases} z, & \text{if } z > 0 \\ 0, & \text{otherwise} \end{cases} \tag{2.6}$$

ReLU is a newly proposed activation function [36], as opposed to all other activation functions that have been used for decades. ReLU is identical to a linear activation function except that ReLU truncates negative output to zero. However, it should be emphasized that ReLU is still a nonlinear activation function.

We now consider building a multilayer feed-forward neural network. A typical way of organizing neurons is by layers. That is to say, we build several layers of neurons, satisfying the following conditions:

- There is no connection of neurons within a layer.
- For each two consecutive layers, each lower-layer node (near the input) is connected to each upper-layer node (near the output), forming a complete bipartite graph.
- There is no connection of neurons between non-consecutive layers.

Such architecture is also known as the multilayer perceptron (MLP), illustrated in Fig. 2.3. The input of data features is called an *input layer*, the prediction (either regression or classification) is called an *output layer*, and others are called *hidden layers*. The community has different methods to count the number of layers in a neural network. In our book, the input is not counted because it is not parametrized. Thus, the example in Fig. 2.3 is a two-layer neural network, which has one hidden layer.

In fact, the above criterion of building neural networks is not essential. Researchers have explored various architectures for feed-forward neural networks. The only

Fig. 2.3 Multilayer
perceptron, which is
essentially a layer-wise stack
of neurons

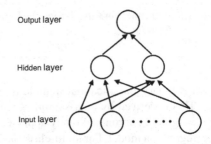

requirement is that the network has to be a *directed acyclic graph* (DAG). This
is also what we mean by *"feed-forward."*

The computation of a feed-forward neural network starts from the input layer,
and then calculates the activation value of each neuron along the propagation path.
Formally, let the input of a layer be $x \in \mathbb{R}^{N_x}$ (which is the output of the consecutive
lower layer), and let the output of the layer be $y \in \mathbb{R}^{N_y}$, where N_x and N_y are the
numbers of input and output neurons, respectively. Then, we have

$$z_i = \sum_{j=0}^{N_x} w_{ij} x_j + b_i \tag{2.7}$$

$$y_i = f(z_i) \tag{2.8}$$

It should be emphasized that in multilayer neural networks, we typically choose
nonlinear activation functions because multiple linear transformations are yet another
linear transformation.

2.1.2 Training Objectives

Nearly all machine learning approaches involve an explicit objective, toward which
we are optimizing the underlying model by training. The *objective function*, denoted
as J in this book, is also called a *"loss function,"* or *"cost function."*

Let us consider supervised learning objectives, and the following are common
loss functions.

- **Mean square error (MSE)**. If the problem at hand is a regression problem, that
  is, the target value t is a real number, it is typical to apply the mean square error,
  defined as

$$J_{\mathrm{MSE}} = \frac{1}{2N} \sum_{i=1}^{N} |y^{(i)} - t^{(i)}|^2 \tag{2.9}$$

where $y^{(i)}$ is the neural prediction and $t^{(i)}$ is the goundtruth target value, respectively, for the ith data point. N is the number of data points.

Although MSE can be interpreted as a heuristic objective, MSE is in fact equivalent to maximum likelihood estimation (MLE) if we assume t is perturbed with Gaussian noise centered at its true value.

- **Cross entropy loss (binary case).** If the task is a binary classification problem where $t \in \{0, 1\}$, cross entropy loss can be applied to the sigmoid output. That is

$$y^{(i)} = \sigma(\boldsymbol{w}^\top \boldsymbol{h}^{(i)} + b) \tag{2.10}$$

$$J_{\text{XENT}} = \frac{1}{N} \sum_{i=1}^{N} \left[-t^{(i)} \log p(y^{(i)}) - (1 - t^{(i)}) \log(1 - y^{(i)}) \right] \tag{2.11}$$

where $\boldsymbol{h}^{(i)} \in \mathbb{R}^{N_h}$ is the last hidden layer of the ith data point. $\boldsymbol{w} \in \mathbb{R}^{N_h}$ and $b \in \mathbb{R}$ are parameters. Notice that MSE is not proper for classification problems because MSE is insensitive when the difference between $y^{(i)}$ and $t^{(i)}$ is small, which, in turn, makes thresholding classification vulnerable. Cross entropy loss is related to MLE of a Bernoulli distribution of y, and is more suited to classification tasks. In Eq. (2.10), the sigmoid function σ comes into place by the assumption of generalized linear distribution on $\boldsymbol{h}^{(i)}$ and $y^{(i)}$.

- **Cross entropy loss (multi-class case).** In this scenario, a classification problem has multiple categories where only one is the groundtruth label of a particular input. That is to say, $t_{\text{id}}^{(i)} \in \{1, 2, \ldots C\}$, where C is the number of categories. t_{id} is called the *index representation*. Equivalently, we can represent the target as a vector, each element indicating whether the data point belongs to this category. Thus, the target is denoted as $t_{\text{onehot}}^{(i)} \in \{0, 1\}^C$, known as the *one-hot representation*.[2] Given an input data point, we assume the target $t^{(i)}$ is a multinomial distribution.[3] The output of the neural network typically uses the softmax function as

$$y_j = \text{softmax}(\boldsymbol{w}_j^\top \boldsymbol{h}^{(i)} + b_j) \overset{\text{def}}{=} \frac{\exp\{\boldsymbol{w}_j^\top \boldsymbol{h}^{(i)} + b_j\}}{\sum_{j'} \exp\{\boldsymbol{w}_{j'}^\top \boldsymbol{h}^{(i)} + b_{j'}\}} \tag{2.12}$$

where $\boldsymbol{w}_j \in \mathbb{R}^{N_h}$ and $b_j \in \mathbb{R}$, for $j = 1, \ldots, C$, are parameters. Now, cross entropy loss for the multi-class case is

$$J_{\text{XENT}} = -\frac{1}{N} \sum_{i=1}^{N} \sum_{j=1}^{C} t_j^{(i)} \log y_j^{(i)} \tag{2.13}$$

---

[2] We can unambiguously distinguish from the font if the target label is represented by the index or one-hot vector. Therefore, it is common to omit the superscripts "id" and "onehot."

[3] The assumption is in fact trivial because every finite, discrete distribution is a multinomial distribution.

Since $t^{(i)}$ is the one-hot representation of the target value, only one term of the inner-summation is non-zero. This is equivalent to

$$J_{\text{XENT}} = -\frac{1}{N} \sum_{i=1}^{N} \log y_{t_{\text{id}}^{(i)}}^{(i)} \tag{2.14}$$

- **Auxiliary loss.** In addition to the losses above, we can apply some auxiliary loss as a part of the objective. For example, the ℓ_2 penalty can be used to regularize neural models [38], given by

$$L_2 = \|\boldsymbol{\theta}\|_2^2 = \sum_i \theta_i^2 \tag{2.15}$$

where $\boldsymbol{\theta}$ is the vector of all parameters, that is to say, matrices are vectorized and all parameters are concatenated. ℓ_2 penalty enforces the model parameters to have a small norm. It is equivalent to do max a posteriori estimation of parameters with a Gaussian prior. In a neural network with sigmoid and tanh activation functions, small parameters ensure that the activation values are around linear region of the sigmoid or tanh function. Thus, the multilayer neural network would behave similarly to a single linear transformation, as multiple linear transformations are yet another linear transformation. When combined with an auxiliary loss, the total objective for a classification task, say, is

$$J = J_{\text{XENT}} + \lambda L_2 \tag{2.16}$$

where λ is a *hyperparamter* that balances the two losses. By "hyperparameter," we mean that λ is specified in advance, as opposed to being learned by the neural network. Hyperparameters can be manually tuned by validation on a hold-out set. That is to say, data samples are divided into three disjoint parts: training, validation, and test sets. We train the model on the training set with different hyperparameters, choose the model with the highest validation performance, and report the test performance.

2.1.3 Learning Neural Parameters

Parameter learning, also known as *model training*, is usually the key of neural networks as well as other machine learning algorithms. Since there is no analytic solution to general neural networks, parameter learning is accomplished by gradient-based methods. The general idea is to update until convergence by

$$\theta_i \leftarrow \theta_i - \alpha \frac{\partial J}{\partial \theta_i}, \quad \forall i \tag{2.17}$$

where α is a hyperparameter known as the *learning rate*. It adjusts the step size of each gradient update. In the following, we first introduce how the gradient $\frac{\partial J}{\partial \theta_i}$ is computed, and then we discuss more about gradient update.

2.1.3.1 Backpropagation

When multilayer neural networks were proposed, people did not know how to compute the gradient efficiently. A naïve approach is numerical computation—for each parameter, we perturb it by a small value ε. According to definition, the gradient is

$$\frac{\partial J}{\partial \theta_i} = \frac{J(\theta_i + \varepsilon) - J(\theta_i - \varepsilon)}{2\varepsilon} \tag{2.18}$$

if $\varepsilon \to 0$. For the computational purpose, we can set ε to a small value (e.g., 0.01) and estimate the gradient numerically. However, such computation is extremely inefficient because a neural network has a huge number of parameters, and the loss function J is computed twice for each parameter by Eq. (2.18).

The backpropagation algorithm computes $\frac{\partial J}{\partial \theta_i}$ analytically. The general idea is to iteratively compute the partial derivative of J with respect to the activation of a layer, which in turn can be used to compute the gradient of parameters. Any textbook involving neural networks will introduce the backpropagation algorithm. However, some textbooks use unnecessary terminologies, e.g., error propagation, Delta rule [4, 15], whereas others use over-complicated notations [14, 55]. These make backpropagation hard to understand.

In our book, we introduce backpropagtion from the view point of chain rule of derivatives. Moreover, we use matrix-vector notations, which not only simplify the derivation but also comply with efficient implementation in high-level programming language (e.g., Matlab, python). The following equations can further serve as the cheat sheet of implementing neural networks.[4]

First, we review the forward-propagation process. For each layer, Eqs. (2.7) and (2.8) can be rewritten as

$$z = Wx + b \tag{2.19}$$

$$y = f(z) \tag{2.20}$$

where $x \in \mathbb{R}^{N_x}$ and $z, y \in \mathbb{R}^{N_y}$. $W \in \mathbb{R}^{N_y \times N_x}$ and $b \in \mathbb{R}^{N_y}$ are the weights and bias term. N_x and N_y are the numbers of neurons in the lower and upper layers, respectively.

---

[4]The backpropagation equations are useful only when we implement backpropagation manually. Nowadays, mature auto-differentiation tools are available, e.g., `TensorFlow` abd `pytorch`, where backpropagation is handled automatically. However, it is still interesting to understand backpropagation from a mathematical perspective, and manual implementation is also a fun exercise.

We now consider backpropagation for gradient computation. As said, backpropagation is a recursive algorithm, and the recursion variable is $\frac{\partial J}{\partial y}$. The recursion process is as follows.

- **Initialization**. The recursive computation starts from the output layer, for which $\frac{\partial J}{\partial y}$ is defined by the loss, viewed as a function of y, i.e., $J(y)$. For MSE, we have $J_{\text{MSE}} = \frac{1}{2}\|y - t\|^2$ for a particular data point, where we omit the superscript (i). Its derivative is $\frac{\partial J}{\partial y} = y - t$. For cross entropy loss, $\frac{\partial J}{\partial y}$ is a little complicated, but $\frac{\partial J}{\partial z}$ has a neat form, given by $\frac{\partial J}{\partial z} = z - t$.

- **Recursion**. We recursively compute the quantity $\frac{\partial J}{\partial y}$ for each layer. Notice that, the lower layer's activation values are the input of the upper layer, denoted as x in our notation. Thus, $\frac{\partial J}{\partial x}$ is the quantity that we need to compute. According to the chain rule of the derivative of compositional functions, we have

$$\frac{\partial J}{\partial z} = \sum_j \frac{\partial J}{\partial y_j} \frac{\partial y_j}{\partial z} = \frac{\partial J}{\partial y} \circ \frac{\partial y}{\partial z} \tag{2.21}$$

$$\frac{\partial J}{\partial x} = W^\top \frac{\partial J}{\partial z} \tag{2.22}$$

In Eq. (2.21), we abuse the notation $\frac{\partial y}{\partial z}$ to denote $\left[\frac{dy_1}{dz_1}, \frac{dy_2}{dz_2}, \ldots, \frac{dy_{N_y}}{dz_{N_y}}\right]^\top$, noticing that y is an element-wise function of z. "$\circ$" is an element-wise product of vectors, also known as the *Hadamard product*. Now, we are ready to compute the derivatives with respect to W and b, given by

$$\frac{\partial J}{\partial W} = \frac{\partial J}{\partial z} x^\top \tag{2.23}$$

$$\frac{\partial J}{\partial b} = W^\top \frac{\partial J}{\partial z} \tag{2.24}$$

It should be also emphasized that the gradients $\frac{J}{\partial W}$ and $\frac{J}{\partial b}$ are obtained by a particular data sample. If we would like to compute the gradient of the total loss as in Eq. (2.16), say, we need to take an average over all data samples. In this case, we group column vectors x, y, etc. of different samples as matrices X, Y, etc., respectively. Then, Eqs. (2.21)–(2.24) take the same form.

- **Termination**. The backpropagation algorithm terminates if it has computed the derivative for all layers. For the input, we do not have to compute Eq. (2.22), as the input features are not a part of the neural network.

It should be noticed that backpropagation is a linear system, which essentially implies that gradients are additive for weight sharing and multiple cost functions. However, the linearity of backpropagation results in the problem of gradient vanishing or exploding. The intuition is that if we multiply the gradient with a large number of matrices during backpropagation, it is likely that the gradient will be either very large or very small.

2.1.3.2 Gradient Descent

Having computed the gradient of the objective with respective to parameters, we can optimize our model with the gradient update (2.17). In practice, we have different variants regarding how many data points to use for an update step.

- **Full-batch gradient descent**. Using the gradient of total loss is the orthodox version of gradient descent. As said, it is inefficient because the training set may contain thousands of or millions of samples. A single step of update involving that much computation is unaffordable. Moreover, full-batch gradient descent may lead to less generalizable solutions [25].
- **Stochastic gradient descent (SGD)**. Stochastic gradient descent picks a random sample at a time, and updates the parameter according to the gradient of the sample loss. SGD is the other extreme as opposed to the full-batch version. The gradient is noisier than full-batch and may need more steps to converge, but it may help neural networks escape from poor local optima.
- **Mini-batch gradient descent**. In this variant, the gradient for each update is computed for the loss summed over a small random subset of training data. The number of samples in the subset is called the *batch size*. The mini-batch version is a compromise of the computation for each gradient update and the number of updates needed for convergence.[5] It is also efficient in GPU computation, supported by most modern neural packages (e.g., Theano, TensorFlow, and PyTorch).

Another recent advance of gradient-based optimization is adaptive gradient descent. In Eq. (2.17), the parameters are updated locally to the point that minimizes the loss in a neighboring region. Thus, it is also known as *steepest gradient descent*. However, the local direction that optimizes the loss may not be the same as the ultimate direction we would like to go. For example, if the contour of the loss is elongated ellipses, the point close the long axis but far from the short axis will have a gradient that is almost orthogonal to the direction to the global optimum. Therefore, steepest gradient descent is inefficient. Momentum methods can be applied for damping in the noisy direction [46]. Other algorithms (e.g., AdaDelta [58] and Adam [27]) compute the learning rate adaptively for each parameter. These approaches have intuitive explanation as well as mathematical justification after series of approximations. Unfortunately, the results in this book were achieved earlier than the wide application or even the invention of these algorithms, and we are not going to introduce these advanced algorithms here. Interested readers are referred to the above papers for details.

---

[5]An interesting terminological abuse is that textbook stochastic gradient descent (SGD) usually refers to updating with a single data point, i.e., the batch size is 1, but that it may refer to mini-batch gradient descent in research papers with a batch size greater than 1. In our book, we follow the convention of the literature and abuse the two terminologies if needed.

2.1.4 Pretraining Neural Networks

In the previous subsection, we present basic algorithms to train a neural network, where "training" is said in the sense of directly optimizing the prediction objective (e.g., classification).

Pretraining is another way of learning neural parameters. However, the objective of pretraining is not the prediction goal. In image classification, for example, training a neural network directly optimizes the classification accuracy, whereas pretraining could be applied before training to learn general features of images, which may or may not be relevant to the particular classification task at hand. Then, the image classification model uses the pretrained parameters as a meaningful initialization. An optional but helpful step, called *fine-tuning*, can be further applied to make parameters more suited to the classification problem. A more profound advantage is that pretraining can make use of massive unlabeled datasets because it learns features in an unsupervised fashion. This could help the performance for a specific task that has not enough labels.

In early years, multilayer neural networks were very difficult to train, and Hinton et al. [18] proposed a pretraining algorithm based on restricted Boltzmann machines (RBMs), which largely improved the training of neural networks at that time. However, such methods become more or less obsolete due to the huge increase of computational resources and huge volume of data available these days. Therefore, we do not cover RBMs in this book.

Another common approach of pretraining is the autoencoder [2]. It is widely applied as an unsupervised method of feature extraction, and inspires our coding criterion for program representation learning (Sect. 4.2.2). Thus, we would like to give a brief introduction as below.

An autoencoder first encodes an input feature x to some hidden representation h, from which it decodes an estimated value of x, denoted as $\hat{x}$. It could be reasonably expected that h captures some "anonymous" features of x, and thus the encoder's weights are a meaningful initialization of downstream tasks, e.g., image classification. Figure 2.4 illustrates the autoencoding process.

Formally, let $x \in \mathbb{R}^{N_x}$ be the input feature and $h \in \mathbb{R}^{N_h}$ be the encoded hidden representation. Here, we assume $N_h < N_x$ holds strictly. The encoder can be realized with a neural network, given by

Fig. 2.4 Illustration of an autoencoder

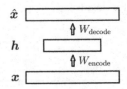

$$h = f_{\text{encode}}(W_{\text{encode}}x + b_{\text{encode}}) \qquad (2.25)$$

where W_{encode} and b_{encode} are the parameters of the encoder.

During the decoding phase, we could use a similar model as

$$\hat{x} = f_{\text{decode}}(W_{\text{decode}}h + b_{\text{decode}}) \qquad (2.26)$$

where W_{decode} and b_{decode} are the parameters of the decoder.

Suppose we are dealing with real-valued features, the objective can be the mean square error defined in Eq. (2.9), and the optimization variables are W's and b's. That is to say, we would like to minimize

$$J(W_{\text{encode}}, b_{\text{encode}}, W_{\text{decode}}, b_{\text{decode}}) = \frac{1}{2}\|\hat{x} - x\|^2 \qquad (2.27)$$

For categorical features x, cross entropy as in (2.11) or (2.13) could be used as the reconstruction loss.

An autoencoder only makes sense when the hidden layer has (far) fewer units than the input. Otherwise, the network might simply learn a trivial encoding. For example, if f_{encode} and f_{decode} are identity functions, any W_{encode} and W_{decode} satisfying $W_{\text{decode}} = W_{\text{encode}}^{-1}$ would give perfect reconstruction. It should also be mentioned that if f_{encode} and f_{decode} are identity, the autoencoder is equivalent to principle component analysis (PCA), except that PCA is the analytic solution with several additional constraints. In this sense, an autoencoder can be intuitively thought of as "nonlinear principle component analysis."

If we train a stack of autoencoders greedily, we could use them to initialize the parameters of a task-specific deep neural network, serving as a way of pretraining. For a stack of autoencoders, it does not make much sense that each subsequent encoding layer has fewer nodes than the previous layer because we cannot learn a wide enough hidden layer if the model is deep. To alleviate this, *denoising autoencoders* [53, 54] perturb x with some noise and obtain $\tilde{x}$, from which the encoder and decoder hope to reconstruct the original data x. With corrupted noise on input, such autoencoder is expected to capture the nature of data. An alternative solution is the *sparse autoencoder* [29], whose training objective contains an axillary sparsity penalty. Thus, sparse autoencoders could learn a hidden layer with more units than input, but only a few are active for a data point.

As has been discussed before, however, pretraining is less interesting than feature extraction itself.

2.2 Neural Networks for Natural Language Processing

In our book, the application domains are natural language processing (NLP) and programming language processing. The latter is a new field, which resembles NLP much, as opposed to image and speech domains where signals are real-valued numbers. Therefore, we would like to pay special attention to the neural models in NLP. Section 2.2.1 addresses the specialty of the NLP domain. Sections 2.2.2 and 2.2.3 introduce neural language modeling and word embeddings, respectively.

2.2.1 Specialty of Natural Language Processing

NLP deals with discrete signals. In image or speech processing, the data are usually some physical quantities, for example, the intensity of a pixel and the frequency of sound. These quantities are continuous variables whose ranges are the real numbers or its continuous subset. In NLP, however, the most widely applied granularity is the word. It can be decomposed of characters and syllables, and can compose phrases, sentences, paragraphs, etc. None of these signals are continuous.

A very naïve and dangerous solution is to treat word indexes as features. Although a large enough neural network can approximate any function [8], it is improper to feed such indexes to neural networks because there is no "semantic" order in these indexes.

An alternative solution is to use one-hot representations as features, i.e., each word is represented as a $|\mathcal{V}|$-dimensional vector, where $|\mathcal{V}|$ is the vocabulary size. For the ith word in the vocabulary, the ith element in the vector is 0, and all others are zero. If we feed such representation to a neural layer with parameters $W \in \mathbb{R}^{N_e \times |\mathcal{V}|}$, it is equivalent to extract the ith column of the word. This shows a one-one correspondence between words and the column vectors of W. Since $N_e \ll |\mathcal{V}|$, the column vectors are thus called *word embeddings*. In NLP, the vocabulary size is usually large, for example, 10k–100k, and therefore, unsupervised pretraining of word embeddings are helpful in various NLP tasks. The training of word embeddings will be discussed within this section.

Natural language contains rich, varying structures. Let us consider an image with 28×28 pixels. The 2D image can be vectorized as a 784-dimensional vector. If each pixel is represented as the gray value in range interval $[0, 1]$, then the image is located in the hypercube $[0, 1]^{784}$. However, the length of a natural language sentence is indefinite, thus data cannot be represented as a vector in a certain dimensional space (even we have considered its discreteness). Moreover, natural language sentences have rich internal structures, for example, constituency parse trees and dependency parse trees. Such information can be explicitly fed to neural networks, or otherwise, the network has to learn an implicit parser for each task. This property is particularly addressed in Sect. 2.3, and is also the main premise of this book.

2.2.2 Neural Language Models

Language modeling treats a natural language corpus as random variables and optimizes the joint probability of the corpus. Although it is controversial if probability is indeed useful for NLP, language modeling is related to various NLP tasks, and has inspired many important NLP models. Therefore, language modeling itself is a hot research topic in NLP. A related application could be speech recognition. If a system recognizing some sound like "*It's a dog.*" The sentence is more likely to be "*It's a dog*" than "*Its a dog.*" A potential reason is that the former has a larger "probability," and thus is more likely to be uttered by humans. However, the probability is actually not necessary, as said. We can learn a scoring function to estimate its "naturalness," which enables us to get rid of the annoying normalization factor in the probability.

Formally, let a corpus have m words, namely, $w_1, \ldots, w_m$. Then, the joint probability $p(w_1, \ldots, w_m)$ can be decomposed as[6]

$$p(w_1, \ldots, w_n) = p(w_1)p(w_2|w_1) \cdots p(w_n|w_1^{m-1}) \qquad (2.28)$$

However, we cannot estimate all the parameters $p(w_i|w_1^{i-1})$ for all i and in the above equations. A common treatment is the Markov assumption

$$p(w_i|w_1^{i-1}) \approx p(w_i|w_{i-n+1}^{i-1}) \qquad (2.29)$$

where we have two assumptions

1. Given previous $n - 1$ words, the next word is independent of words before the $n - 1$ words.
2. Given previous $n - 1$ words, the next word is independent of its position in the corpus.

This is called an *n-gram model*. If we consider the conditional distribution of a word $p(w_i|w_{i-n+1}^{i-1})$ as a multinomial distribution, we could estimate its parameters by maximum likelihood estimation, which is essentially counting the occurrence

$$\hat{p}(w_n|w_1^{n-1}) = \frac{\# \text{ of } w_1^n}{\# \text{ of } w_1^{n-1}} \qquad (2.30)$$

In such language model, we have $|\mathcal{V}|^{n-1}$ multinomial distributions because, for any particular combination of w_{i-n+1}^{i-1}, we have a different conditional distribution for w_n with $|\mathcal{V}|$ parameters. In total, we have $\mathcal{O}(|\mathcal{V}|^n)$ parameters.

Since natural language words usually follow the power-law distribution [32]: a large number of words only appear a few times, whereas a few words appear a large number of times. Therefore, the n-gram model suffers from the data sparseness problem even for very small n (e.g., $n = 2$ or 3). Once a particular word combination in the test set has never occurred in the training set, its conditional probability will be estimated as zero, which in turn makes the probability of the entire corpus zero, as indicated in Eq. (2.28). This has a disastrous effect on the language model. Various smoothing techniques have been proposed to alleviate the problem. We do not contain the details in this book due to the topic constraints, and refer interested readers to [23].

---

[6]We denote $w_i, w_{i+1}, \ldots, w_j$ by w_i^j for short.

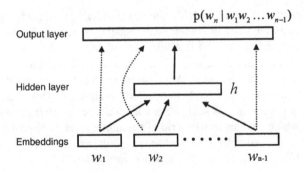

Fig. 2.5 A neural language model (adapted from [1]). The embeddings are concatenated, and then fed to a hidden layer. There are also bypassing connections (dotted lines) from embedding layers to the output layer, which are more about a design choice to improve performance than the key of the model

Neural language models differ from n-gram models in that, instead of modeling $|\mathcal{V}|^{n-1}$ probabilistic distributions, a neural language model only parametrizes a single distribution of the next word, given previous words as input. The earliest neural language model dated back to [1], whose architecture is shown in Fig. 2.5.

The model also takes the Markov assumption (similar to the n-gram model), but designs a feed-forward neural network to predict the next word with a softmax layer, given previous $n - 1$ words.

In the input layer, each word is first represented as an embedding vector, resulting $n_e \cdot |\mathcal{V}|$ parameters for the entire vocabulary; the embedding-to-hidden weight matrix contains $n \cdot n_e \cdot n_h$ parameters; and the output layer contains $n_h \cdot |\mathcal{V}|$. In this way, the number of parameters grows linearly with respect to n, as opposed to exponential growth in the naïve n-gram model.

An extension of the feed-forward neural language model is to use a recurrent neural network to capture all previous words from w_1 to w_{i-1} [33]. Due to the organization of this book, the recurrent neural network will be introduced in Sect. 2.3.2.

2.2.3 Word Embeddings

A byproduct of a neural language model is the learned vector for each word (also known as word *embeddings*), i.e., the bottom vectors in Fig. 2.5. These embeddings can be intuitively thought of as a weight matrix applied to the one-hot representation of a word, illustrated in Fig. 2.6. Since exactly one element in the one-hot vector is on, the matrix-vector product retrieves one column in the matrix, and does not mix other columns. Therefore, a direct word-to-vector mapping is preferred than such matrix-vector multiplication due to implementation efficiency.

For the purpose of word embedding learning, it is unnecessary to deal with the normalization factor in the softmax output layer of the language model. Therefore, various studies propose more dedicated models for word embedding learning. We would like to introduce a few in the following.

Fig. 2.6 An interpretation of word embeddings

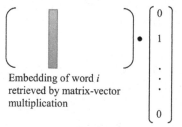

Embedding of word i retrieved by matrix-vector multiplication

One-hot representation of word i (sparse)

C&W Model. Collobert and Western [7] propose a unified neural model for part-of-speech tagging, named entity recognition, semantic role labeling, etc. In their work, they also propose a method of word embedding learning based on hinge loss and negative sampling.

Given a subsequence of word $w = w_1, \ldots, w_n$, we train a scoring function $f(w) \in \mathbb{R}$ to indicate the "naturalness" of w, instead of its probability. To prevent the network from learning a trivial scoring function that assign all candidates equally large scores, we need some *negative samples*, denoted as w^-, to pull down the scoring function during the training process. In particular, their training objective is to hope $f(w)$ to be greater than $f(w^-)$ by a constant Δ. In general, the training objective is

$$J = \max\{0, \Delta - f(w) + f(w^-)\} \tag{2.31}$$

Such objective is similar to the support vector machine (SVM), and called the *hinge loss*. Our coding criterion in Sect. 4.2.2 uses the same objective to train program vector representations.

In the negative sampling schema, the probabilistic distribution and the number of negative samples are among the most important hyperparameters. Generally, we would like to have more negative samples closer in distribution to positive samples. A simple intuition is such that, if the negative samples are very different from positive samples, the discrimination of positive and negative samples is simple, resulting in poor embedding learning. Readers can refer to [13] for detailed discussion on negative samples and the noise-contrastive estimation approach.

CBoW and Skip-gram models. Pushing the complexity of embedding learning to the minimum are Mikolov et al.'s continuous bag-of-words and skip-gram models [34]. They use a context vector w_I, also known as an input vector, to predict a word v with softmax weight v_O,[7] given by

---

[7]Subscripts I and O represent *input* and *output*, respectively.

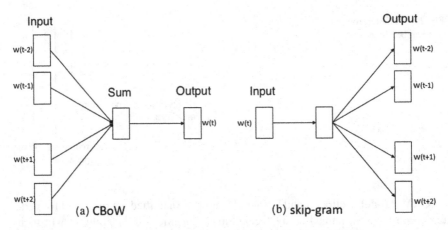

Fig. 2.7 **a** CBoW model; and **b** skip-gram model. (Figure is adapted from [34])

$$p(v|\cdot) = \frac{\exp\{\boldsymbol{v}_O^\top \boldsymbol{w}_I\}}{\sum_{u \in \mathcal{V}} \exp\{\boldsymbol{u}_O^\top \boldsymbol{w}_I\}} \qquad (2.32)$$

Such a particular model is known as `word2vec`.

Further, they have two variants of modeling $\boldsymbol{w}_I$. The continuous bag-of-words (CBoW) defines a window and takes the average of context words' intput embeddings as $\boldsymbol{w}_I$. The skip-gram uses a single word to predict surrounding words. These variants are illustrated in Fig. 2.7. To alleviate the complexity issue in the softmax, they use either hierarchical softmax or noise-contrastive estimation to approximate Eq. (2.32).

Word embedding learning can be thought of as a pretraining method because the training objective is not task-related. Unlike the restricted Boltzmann machine, which has been obsolete these days, word embeddings are widely applied in NLP. We would like to point out several typical usages.

- **Initialization of neural networks.** Word embeddings can be helpful for various NLP tasks [16, 22, 31, 35, 43, 44, 49, 51, 59]; they are also useful as features for traditional machine learning models, for example, the conditional random field (CRF) [12]. However, in recent NLP tasks with large-scale datasets, e.g., machine translation, pretrained word embeddings may not be useful any longer; they can be simply randomly initialized and learned by backpropagation.
- **Similarity measuring.** The other application of word embeddings is the similarity measure. The Euclidean distance or cosine distance of two word embeddings reflects the similarity of the two words. This is often used as a way of evaluating word embedding learning approaches. The embedding-based similarity measure can be extended to the sentence level by heuristics, for example, averaging word embeddings. It is useful for obtaining a matching model in a no-data scenario [45].
- **Linearity of word relations.** Despite general similarity of words, word embeddings learned by a neural model have an intriguing property: the relationship between a pair of words can often be measured by the linear offset of embeddings,

for example "man"–"woman" $\approx$ "king"–"queen." Such property can be observed for both syntactic (e.g., tense) and semantic relations; some researchers leverage such property to learn relationship between concepts [9].

In addition to the embeddings of words, such idea can be applied in various domains where the input signal is discrete, e.g., the social network [39], knowledge base [5], and e-Commerce products [11]. Our programs' vector representation in Sect. 4.2.2 is also in this line. All these embeddings provide a way of learning anonymous features of discrete input, and map them into low-dimensional space.

2.3 Structure-Sensitive Neural Networks

Section 2.1 introduce a generic multilayer neural network, whose input is a fix-sized vector. In this section, we discuss structure-sensitive neural networks, including convolutional neural networks (Sect. 2.3.1), recurrent neural networks (Sect. 2.3.2), and recursive neural networks (Sect. 2.3.3).

2.3.1 Convolutional Neural Network

Convolutional neural networks (CNNs) were initially designed for images [30], among the few successful neural networks in early years. Different from a layer-wise fully connected neural network, CNN uses a small sliding window to extract local features and then aggregates these features by pooling. Therefore, CNN reduces the number of parameters to a large extent, but can effectively extract features over different regions of an image. Further, such sliding windows are translation (moving) invariant. This is actually a desired property of image processing because if we would like to detect a cat, for example, we do not care where the cat is located in the image.

Let $f : \mathbb{R}^2 \to \mathbb{R}$ and $k : \mathbb{R}^2 \to \mathbb{R}$ be 2D signals. From a signal processing perspective [21], the *convolution* of f and k, denoted as $f * * k$, is

$$(f * * k)(x_1, x_2) = \int_{-\infty}^{\infty} \int_{-\infty}^{\infty} f(\tau_1, \tau_2) k(x_1 - \tau_1, x_2 - \tau_2) \mathrm{d}\tau_1 \mathrm{d}\tau_2 \qquad (2.33)$$

for the continuous case, or

$$(f * * k)(x_1, x_2) = \sum_{\tau_1 = -\infty}^{\infty} \sum_{\tau_2 = -\infty}^{\infty} f(\tau_1, \tau_2) k(x_1 - \tau_1, x_2 - \tau_2) \qquad (2.34)$$

for the discrete case.

Fig. 2.8 A convolutional
neural network (CNN),
which typically comprises an
embedding layer, a
convolutional layer, a
pooling layer, and an output.
Reprinted from [35] with
permission. © 2015,
Association for
Computational Linguistics

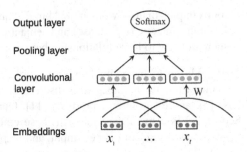

Intuitively, the convolution operator first flips the k function (also called a *convolution kernel* or a *filter*), and then operates "inner-product" in the functional space of f and g. If we consider the discrete case with a finite support kernel, the RHS of Eq. (2.34) is essentially the inner-product of vectorized/munrolled f and k, evaluated at a certain point specified by (x_1, x_2). Noticing that $f * *g : \mathbb{R}^2 \to \mathbb{R}$ is itself a 2D function, we should evaluate the inner-product in the above step at different values of (x_1, x_2). In summary, the convolution operator works in a "flip-product-slide" fashion. In traditional signal processing, convolution with a Gaussian kernel has a blurring effect, whereas a Laplacian kernel has a sharpening effect [48].

If we would like to learn the appropriate kernel(s) for image processing with machine learning methods, the kernel k should be discrete with a finite support, where the values are parameters. In this case, however, the "flipping" step is unnecessary because the training process could directly learn a flipped kernel. The inner-product coincides with the linear transformation of a neuron as in Eq. (2.1). If we learn a set of kernels, it is equivalent to have a layer of neurons.

Moreover, there is nothing preventing from adding a bias term and using a non-linear activation function for the neural network, although the orthodox convolution in the signal processing is a linear operator. The sliding step then becomes the key of CNN, that is, CNN is essentially a locally connected layer, sliding over the entire input.

Due to the main topic of this book, we now consider a CNN for NLP, shown in Fig. 2.8. Typically, it takes word embeddings as input, and then a fixed-size window extracts local features of words, with a subsequent pooling layer aggregating information over different regions of the sentence. Finally, an output layer, for example, the softmax layer for classification, predicts the label.

Formally, let t be the size of the convolution window. Suppose the window is at the position of $w_1, \ldots, w_t$, whose embeddings are $\boldsymbol{w}_1, \ldots, \boldsymbol{w}_t \in \mathbb{R}^{n_e}$. Then the output of the convolution, evaluated at the current words, is a vector $\boldsymbol{y} \in \mathbb{R}^{n_c}$, where n_c is the number of convolutional kernels/filters, or sometimes known as the dimension of convolution. This process is given by

$$y = f(W[\boldsymbol{w}_1, \ldots, \boldsymbol{w}_t] + b) \tag{2.35}$$

where $W \in \mathbb{R}^{n_c \times (t \cdot n_e)}$ and $\boldsymbol{b} \in \mathbb{R}^{n_c}$ are parameters, and f is the activation function.

When the convolution kernel moves to the boundary of input signal (e.g., the beginning and ending position of the sentence), it raises a problem that there are not enough words in the convolution kernel. Some studies assign a special token indicating the outside of input boundary [24], while others pad the sentence with zero vectors [20, 38].

There are more tricks about CNNs. In image processing, sliding can be conducted several pixels at the time, known as *striding*. This is because nearing pixels tend to have the same intensity; thus every-pixel sliding is unnecessary, but striding helps to eliminate features to a large extent. However, such striding is not applied in NLP because, in a natural language sentence *"the cat sat on the mat,"* for example, the consecutive positions of a three-word window, *"the cat sat"* and *"cat sat on,"* have very different meanings. By contrast, multi-size windows [26] are used for CNNs in NLP, analogous to n-gram models with different values of n.

Let $y_1, \ldots, y_t$ be the features extracted by convolution at different words. If we would like it to be processed by a softmax layer for prediction, we have to compress these vectors to a fixed-size vector, denoted by p here. A common approach is *max-pooling*, that is, choosing the maximum value in each dimension, given by

$$p[i] = \max \left\{ y_1[i], y_2[i], \ldots, y_t[i] \right\} \quad \forall i \tag{2.36}$$

Sum pooling, average pooling, and min pooling can also be applied. The details are not repeated. Pooling in NLP is also different from image processing. Images usually have a large number of pixels, and thus pooling can reduce the dimension of signals by taking the maximum value of several neighboring convolution features. However, there are far fewer words in a natural language sentence, but the length of a sentence may vary for different data points. Thus, pooling in NLP works in a *dynamic* fashion: the max operator in (2.36) is applied to a varying set for different sentences.

An advantage of convolutional neural networks is that the propagation path depends on the architecture design, but not on the size of input. Therefore, it is less sensitive to gradient exploding or vanishing. However, due to the sliding window mechanism, the extracted features are local, which means that two words would not have interaction if they do not appear in a same window. Multilayer CNN alleviates the problem by allowing deeper information interaction [24].

It should be mentioned that CNN is a structure-sensitive neural architecture as the sliding window defines neighborhood of pixels. Consider a 28×28-pixel image. The input features for a fully-connected neural network are a 784-dimensional vector. If we randomly shuffle (change the order of) these 784 features before training, the network is exactly the same because there is no order among different input dimensions. However, CNN's sliding window extracts features of physically neighboring pixels. If pixels are shuffled, the performance would be poor.

2.3.2 Recurrent Neural Network

The recurrent neural network (RNN) is an architecture that has loops in its connections, and Fig. 2.9a shows a simple recurrent neural network. For RNN, we have
to redefine its computation, because it does not make sense to propagate information indefinitely along the loop. Analogous to combinational circuits and sequential
circuits, we introduce a discrete time signal to RNN and let RNN propagate information along one edge at a time. In this way, the loopy connection of RNN is defined
between different time steps.

After we unroll an RNN along the time axis, the RNN is feed-foward without any
loop, shown in Fig. 2.9b. However, RNN is still very different from a feed-forward
neural network, because its transition function is the same for all steps, as opposed to
a feed-forward neural network where each layer can learn its own parameters. Thus,
RNN may exhibit chaotic behaviors [52] and is much more difficult to train [3, 37].

RNN is suitable for modeling time-series data (e.g., speech [10]) by its iterative
nature. In NLP, RNN also has wide applications especially for machine translation [47], summarization [50], and other text generation tasks.

Let us consider the RNN in Fig. 2.9. It has a hidden layer h, changing along the
time. A vanilla RNN has a perceptron-like hidden neurons, that is, the state at time
step t, denoted as h_t, is a linear combination of the last step's hidden state h_{t-1} and
the current input x_t, further processed by a nonlinear function f. Formally, we have

$$h_t = f(W_{in}x_t + W_{rec}h_{t-1} + b_h) \tag{2.37}$$

where W_{in} is the weights for input-hidden connections, and W_{rec} is the weights for
hidden-hidden recurrent connections. The input and output highly depend on the task
at hand, and we discuss the following scenarios.

- **Sentence-level classification task**. The input is the word embeddings of a sentence. The classification output can be applied at the last step's hidden state. Alter-

Fig. 2.9 A recurrent neural
network (RNN). **a**
Compressed representation.
b An unrolled representation

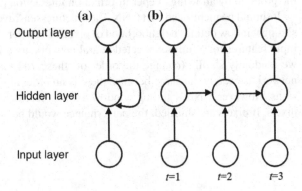

natively, we can introduce a pooling layer on all hidden states, and softmax classification is applied on top of the pooling layer.

- **Sequential labeling**. For tasks like part-of-speech tagging, each word has a label (e.g., "noun" and "verb"), and thus the prediction could be applied to every step. To integrate global information, we can apply a backward RNN starting from the last word and propagating to the first word. Combining with the forward RNN, we have a bidirectional RNN. In addition, the output layer can be easily incorporated with a conditional random field (CRF) for global inference.

- **Sequence generation**. If we would like to generate a sentence with RNN, it becomes tricky for the input layer because we do not have the sentence to be generated. The sequence-to-sequence model [47] addresses the problem as follows. When predicting, the RNN is fed with the last step's generated word, and its output predicts the next word. During training, the generated words are of poor quality, so the ground truth word of the last step is fed into RNN as input.

As said, the vanilla RNN as in Eq. (2.37) has poor performance and is difficult to train because

- The long propagation path makes information imprecise during propagation;
- The gradient exploding or vanishing problem makes it difficult to learn long-term dependency by training; and
- RNN is potentially chaotic because it is a weight-shared iterative nonlinear system.

Researchers have made efforts to improve the dynamics of RNN iteration. A major innovation is the long short term memory (LSTM) units [19]. It was originally proposed in the 1990s by Hochereiter and Schmidhuber, and has attracted renewed interest recently.

The key insight of LSTM is to introduce gating units to balance the current input and previous states. Particularly, the gates are self-adaptive: they determine their gating values based on the current state and input, and are parametrized and learned as other parts of the neural network. Here, we introduce the LSTM variant in [57]. In particular, LSTM has three gating vectors, namely, the input gate i_t, forget gate f_t, and output gate o_t, computed based on the input and the last hidden state and squashed by a sigmoid layer. A candidate feature vector g_t is computed in a similar way, but with a tanh activation function, so that the candidate features are in the range $(-1, 1)$. These temporary variables are computed as

$$i_t = \sigma(W_i x_t + U_i h_{t-1} + b_i) \tag{2.38}$$

$$f_t = \sigma(W_f x_t + U_f h_{t-1} + b_f) \tag{2.39}$$

$$o_t = \sigma(W_o x_t + U_o h_{t-1} + b_o) \tag{2.40}$$

$$g_t = \tanh(W_g x_t + U_g h_{t-1} + b_g) \tag{2.41}$$

where W's and U's are weight matrices, and b's are bias terms.

Then, a vector c_t, called the *memory cell*, is a combination of the candidate feature and the last step's cell, which are weighted by the input gate and the forget gate, respectively. This is given by

Fig. 2.10 The computation
of LSTM units. Reprinted
from [56] with permission. ©
2015, Association for
Computational Linguistics

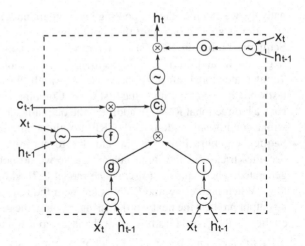

$$c_t = i_t \circ g_t + f_t \circ c_{t-1} \qquad (2.42)$$

And finally the output is weighted by an output gate

$$h_t = o_t \circ \tanh(c_t) \qquad (2.43)$$

The entire process in shown in Fig. 2.10. In recent years, researchers propose other variants, e.g., gated recurrent units (GRUs) [6], to simplify the computation, but they all have the same spirit, with minor difference in computation.

2.3.3 Recursive Neural Network

CNN and RNN are structure-sensitive, as they are aware of the word order in a sentence, as opposed to bag-of-words features. However, they do not consider richer structures of sentences, for example, parse trees. Then, a natural question is: *Is the parse structure useful for neural architectures?*

Pinker illustrates an interesting example in his book, *The Language Instinct*:

The dog the stick the fire burned beat bit the cat.

The sentence is extremely hard to understand by humans due to nested attribute clauses: *"the dog (that) the stick ...beat"* and*"the stick (that) the fire burned."* On the other hand, if the parse tree of the sentence is available as in Fig. 2.11, the meaning of the sentence is clear and can be directly read from the parse tree.

Similar philosophy applies to the neural architecture of sentence modeling. If we do not explicitly use such tree structures, what a CNN sees is the windows like *"the dog the stick," "dog the stick the,"* ..., *"beat bit the cat,"* none of which makes any

Fig. 2.11 The parse tree corresponding to the sentence "*The dog the stick the fire burned beat bit the cat.*"

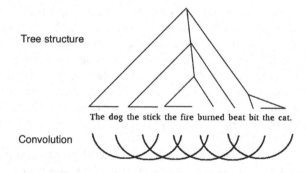

sense. If we use an RNN, it has to learn an implicit parser which could be difficult, especially when the training set is small. With external parse trees, the network can focus more on the task at hand (for example, sentiment analysis).

Socher et al. [41, 43, 44] propose a recursive neural network whose skeleton is a the parse tree. Its intermediate nodes recursively propagate information bottom up, and the root node represents the entire tree, which can be further used for prediction.

Figure 2.12 shows an example of the recursive neural network built upon a constituency parse tree. In the constituency parse tree, each leaf node is a word in the sentence, and the intermediate nodes are abstract components (e.g., a noun phrase or a verb phrase). The root node represents the entire sentence. Let p the parent node with children c_1 and c_2, and their representations are $p, c_1, c_2 \in \mathbb{R}^{n_e}$, respectively. A vanilla recursive neural network has a perceptron-like propagation as

$$p = f(W[x_1; x_2] + b) \tag{2.44}$$

where $W \in \mathbb{R}^{N_e \times 2N_e}$ and $b \in \mathbb{R}^{N_e}$ are parameters.

It should be noted that the dimension of p is the same as that of the child nodes. Moreover, both p and c_1, c_2 are in the same semantic space. This is because the semantics of a vector is determined by how it is used. In the recursive process, all vectors are processed by parameters W and b, and thus, all nodes in a recursive neural network are in the same semantic space.

Fig. 2.12 A recursive neural network, proposed in [43]. Reprinted from [35] with permission. © 2015, Association for Computational Linguistics

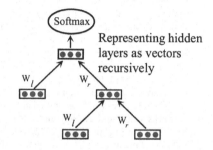

Socher et al. extend their own vanilla recursive neural network with matrix-vector interaction [41] and tensor interaction [44]. These approaches increase the model capacity of the recursive neural network, but do not have significant improvement. Later, with the prospective of LSTM units in recurrent neural networks, several researchers, including Socher et al. themselves, propose to use LSTM interaction for the recursive network [28, 49, 59], achieving much better performance than the vanilla recursive network. In other work, the recursive neural network is extended to the skeleton of dependency parse trees [42] or the combinatory categorical grammar (CCG) [17].

In short, the recursive network can explicitly make use of external tree structures, and learn vectors of nodes recursively bottom up along the tree. The disadvantage of the recursive neural network is similar to recurrent one: if the propagation path is too long, it may lose information and suffer from the difficulty of training.

2.4 Summary

In this chapter, we introduce the basics of a generic neural network, and its training and pretraining methods. Then, we address the application of neural networks in natural language processing (NLP) by pointing out several special properties of NLP; we introduce two NLP-specific neural networks: the neural language model and word embeddings. Finally, we introduce three structure-sensitive neural architectures, namely, convolutional neural networks, recurrent neural networks, and recursive neural networks. These inspire our tree-based convolution, as will be discussed in detail in the next four chapters.

References

1. Bengio, Y., Ducharme, R., Vincent, P., Janvin, C.: A neural probabilistic language model. J. Mach. Learn. Res. **3**, 1137–1155 (2003)
2. Bengio, Y., Lamblin, P., Popovici, D., Larochelle, H.: Greedy layer-wise training of deep networks. In: Advances in Neural Information Processing Systems, pp. 153–160
3. Bengio, Y., Simard, P., Frasconi, P.: Learning long-term dependencies with gradient descent is difficult. IEEE Trans. Neural Netw. **5**(2), 157–166 (1994)
4. Bishop, C.M.: Pattern Recognition and Machine Learning. Springer (2006)
5. Bordes, A., Weston, J., Collobert, R., Bengio, Y.: Learning structured embeddings of knowledge bases. In: Proceedings of the 25th AAAI Conference on Artificial Intelligence, pp. 301–306 (2011)
6. Cho, K., van Merriënboer, B., Bahdanau, D., Bengio, Y.: On the properties of neural machine translation: Encoder-decoder approaches (2014). arXiv preprint arXiv:1409.1259
7. Collobert, R., Weston, J.: A unified architecture for natural language processing: Deep neural networks with multitask learning. In: Proceedings of the 25th International Conference on Machine Learning, pp. 160–167 (2008)
8. Cybenko, G.: Approximation by superpositions of a sigmoidal function. Math. Control Signal Syst. **2**(4), 303–314 (1989)

9. Fu, R., Guo, J., Qin, B., Che, W., Wang, H., Liu, T.: Learning semantic hierarchies via word embeddings. In: Proceedings of the 52nd Annual Meeting of the Association for Computational Linguistics, pp. 1199–1209 (2014)

10. Graves, A., Mohamed, A., Hinton, G.: Speech recognition with deep recurrent neural networks. In: Proceedings of the 2013 IEEE International Conference on Acoustics, Speech and Signal Processing, pp. 6645–6649 (2013)

11. Grbovic, M., Radosavljevic, V., Djuric, N., Bhamidipati, N., Savla, J., Bhagwan, V., Sharp, D.: E-commerce in your inbox: Product recommendations at scale. In: Proceedings of the 21th ACM SIGKDD International Conference on Knowledge Discovery and Data Mining, pp. 1809–1818 (2015)

12. Guo, J., Che, W., Wang, H., Liu, T.: Revisiting embedding features for simple semi-supervised learning. In: Proceedings of Conference on Empirical Methods in Natural Language Processing, pp. 110–120 (2014)

13. Gutmann, M., Hyvärinen, A.: Noise-contrastive estimation of unnormalized statistical models, with applications to natural image statistics. J. Mach. Learn. Res. 13(1), 307–361 (2012)

14. Hastie, T., Tibshirani, R., Friedman, J., Franklin, J.: The Elements of Statistical Learning: Data Mining, Inference and Prediction. Springer (2009)

15. Haykin, S.S., Haykin, S.S., Haykin, S.S., Haykin, S.S.: Neural Networks and Learning Machines. Pearson Education (2009)

16. He, H., Gimpel, K., Lin, J.: Multi-perspective sentence similarity modeling with convolutional neural networks. In: Proceedings of the 2015 Conference on Empirical Methods in Natural Language Processing, pp. 1576–1586 (2015)

17. Hermann, K., Blunsom, P.: The role of syntax in vector space models of compositional semantics. In: Proceedings of the 51st Annual Meeting of the Association for Computational Linguistics, pp. 894–904 (2013)

18. Hinton, G., Osindero, S., Teh, Y.: A fast learning algorithm for deep belief nets. Neural Comput. 18(7), 1527–1554 (2006)

19. Hochreiter, S., Schmidhuber, J.: Long short-term memory. Neural Comput. 9(8), 1735–1780 (1997)

20. Hu, B., Lu, Z., Li, H., Chen, Q.: Convolutional neural network architectures for matching natural language sentences. In: Advances in Neural Information Processing Systems, pp. 2042–2050 (2014)

21. Proakis, J.G., Manolakis, D.G.: Digital Signal Processing: Principles, Algorithms, and Applications. Pentice Hall (1996)

22. Ji, Y., Eisenstein, J.: One vector is not enough: Entity-augmented distributed semantics for discourse relations. Trans. Assoc. Comput. Linguist. 3, 329–344 (2015)

23. Jurafsky, D., Martin, J.: Speech and Language Processing. Pearson Education (2000)

24. Kalchbrenner, N., Grefenstette, E., Blunsom, P.: A convolutional neural network for modelling sentences. In: Proceedings of the 52nd Annual Meeting of the Association for Computational Linguistics, pp. 655–665 (2014)

25. Keskar, N.S., Mudigere, D., Nocedal, J., Smelyanskiy, M., Tang, P.T.P.: On large-batch training for deep learning: Generalization gap and sharp minima. In: Proceedings fo the International Conference on Learning Representations (2017)

26. Kim, Y.: Convolutional neural networks for sentence classification. In: Proceedings of the 2014 Conference on Empirical Methods in Natural Language Processing, pp. 1746–1751 (2014)

27. Kingma, D.P., Ba, J.: Adam: A method for stochastic optimization. In: Proceedings of the International Conference on Learning Representations (2015)

28. Le, P., Zuidema, W.: Compositional distributional semantics with long short term memory (2015). arXiv preprint arXiv:1503.02510

29. Le, Q.V.: Building high-level features using large scale unsupervised learning. In: Proceedings of the 2013 IEEE International Conference on Acoustics, Speech and Signal Processing, pp. 8595–8598 (2013)

30. LeCun, Y., Jackel, L., Bottou, L., Brunot, A., Cortes, C., Denker, J., Drucker, H., Guyon, I., Muller, U., Sackinger, E., et al.: Comparison of learning algorithms for handwritten digit

recognition. In: Proceedings of the International Conference on Artificial Neural Networks, pp. 53–60 (1995)

31. Lei, T., Barzilay, R., Jaakkola, T.: Molding CNNs for text: Non-linear, non-consecutive convolutions. In: Proceedings of the 2015 Conference on Empirical Methods in Natural Language Processing, pp. 1565–1575 (2015)

32. Li, W.: Random texts exhibit Zipf's-law-like word frequency distribution. IEEE Trans. Inf. Theory **38**(6), 1842–1845 (1992)

33. Mikolov, T., Karafiát, M., Burget, L., Cernockỳ, J., Khudanpur, S.: Recurrent neural network based language model. In: Proceedings ot the 11th Annual Conference of the International Speech Communication Association, pp. 1045–1048 (2010)

34. Mikolov, T., Sutskever, I., Chen, K., Corrado, G., Dean, J.: Distributed representations of words and phrases and their compositionality. In: Advances in Neural Information Processing Systems (2013)

35. Mou, L., Peng, H., Li, G., Xu, Y., Zhang, L., Jin, Z.: Discriminative neural sentence modeling by tree-based convolution. In: Proceedings of the 2015 Conference on Empirical Methods in Natural Language Processing, pp. 2315–2325 (2015)

36. Nair, V., Hinton, G.E.: Rectified linear units improve restricted boltzmann machines. In: Proceedings of the 27th International Conference on Machine Learning, pp. 807–814 (2010)

37. Pascanu, R., Mikolov, T., Bengio, Y.: On the difficulty of training recurrent neural networks (2012). arXiv preprint arXiv:1211.5063

38. Peng, H., Mou, L., Li, G., Chen, Y., Lu, Y., Jin, Z.: A comparative study on regularization strategies for embedding-based neural networks. In: Proceedings of the 2015 Conference on Empirical Methods in Natural Language Processing, pp. 2106–2111 (2015)

39. Perozzi, B., Al-Rfou, R., Skiena, S.: DeepWalk: Online learning of social representations. In: Proceedings of the 20th ACM SIGKDD International Conference on Knowledge Discovery and Data Mining, pp. 701–710 (2014)

40. Rosenblatt, F.: The perceptron: A probabilistic model for information storage and organization in the brain. Psychol. Rev. **65**(6), 386 (1958)

41. Socher, R., Huval, B., Manning, C., Ng, A.: Semantic compositionality through recursive matrix-vector spaces. In: Proceedings of the 2012 Joint Conference on Empirical Methods in Natural Language Processing and Computational Natural Language Learning, pp. 1201–1211 (2012)

42. Socher, R., Karpathy, A., Le, Q., Manning, C., Ng, A.: Grounded compositional semantics for finding and describing images with sentences. Trans. Assoc. Comput. Linguist. **2**, 207–218 (2014)

43. Socher, R., Pennington, J., Huang, E., Ng, A., Manning, C.: Semi-supervised recursive autoencoders for predicting sentiment distributions. In: Proceedings of the Conference on Empirical Methods in Natural Language Processing, pp. 151–161 (2011)

44. Socher, R., Perelygin, A., Wu, J., Chuang, J., Manning, C., Ng, A., Potts, C.: Recursive deep models for semantic compositionality over a sentiment treebank. In: Proceedings of Conference on Empirical Methods in Natural Language Processing, pp. 1631–1642 (2013)

45. Song, Y., Mou, L., Yan, R., Yi, L., Zhu, Z., Hu, X., Zhang, M.: Dialogue session segmentation by embedding-enhanced texttiling. Interspeech **2016**, 2706–2710 (2016)

46. Sutskever, I., Martens, J., Dahl, G., Hinton, G.: On the importance of initialization and momentum in deep learning. In: Proceedings of the 30th International Conference on Machine Learning, pp. 1139–1147 (2013)

47. Sutskever, I., Vinyals, O., Le, Q.V.: Sequence to sequence learning with neural networks. In: Advances in Neural Information Processing Systems, pp. 3104–3112 (2014)

48. Szeliski, R.: Computer Vision: Algorithms and Applications. Springer Science & Business Media (2010)

49. Tai, K., Socher, R., Manning, D.: Improved semantic representations from tree-structured long short-term memory networks. In: Proceedings of the 53rd Annual Meeting of the Association for Computational Linguistics, pp. 1556–1566 (2015)

50. Tan, J., Wan, X., Xiao, J.: Abstractive document summarization with a graph-based attentional neural model. In: Proceedings of the 55th Annual Meeting of the Association for Computational Linguistics, pp. 1171–1181 (2017)
51. Tang, D., Qin, B., Liu, T.: Document modeling with gated recurrent neural network for sentiment classification. In: Proceedings of the 2015 Conference on Empirical Methods in Natural Language Processing, pp. 1422–1432 (2015)
52. Thomas Laurent, J.v.B.: A recurrent neural network without chaos. In: Proceedings of the International Conference on Learning Representations (2017). https://openreview.net/forum?id=S1dIzvclg
53. Vincent, P.: A connection between score matching and denoising autoencoders. Neural Comput. **23**(7), 1661–1674 (2011)
54. Vincent, P., Larochelle, H., Bengio, Y., Manzagol, P.: Extracting and composing robust features with denoising autoencoders. In: Proceedings of the 25th International Conference on Machine Learning, pp. 1096–1103 (2008)
55. Webb, A.: Statistical Pattern Recognition. Wiley (2003)
56. Xu, Y., Mou, L., Li, G., Chen, Y., Peng, H., Jin, Z.: Classifying relations via long short term memory networks along shortest dependency paths. In: Proceedings of the 2015 Conference on Empirical Methods in Natural Language Processing, pp. 1785–1794 (2015)
57. Zaremba, W., Sutskever, I.: Learning to execute (2014). arXiv preprint arXiv:1410.4615
58. Zeiler, M.D.: AdaDelta: An adaptive learning rate method (2012). arXiv preprint arXiv:1212.5701
59. Zhu, X., Sobhani, P., Guo, Y.: Long short-term memory over tree structures. In: Proceedings of the 32nd International Conference on Machine Learning, pp. 1604–1612 (2015)

Chapter 3
General Framework of Tree-Based Convolutional Neural Networks (TBCNNs)

Abstract In this chapter, we introduce the general framework of the tree-based convolutional neural network (TBCNN). We first present the design philosophy and the general formula of TBCNN. Then we introduce several applications of TBCNN that will be analyzed in this book. We also highlight the technical difficulties of designing TBCNN in different scenarios, which will be addressed in future chapters.

Keywords Convolutional neural network · Tree-based convolution

3.1 General Idea and Formula of TBCNN

In traditional convolutional neural networks (CNNs), the convolution process can be viewed as a sliding window that extracts local features of data. To process a 28×28-pixel image, for example, CNN applies a set of sliding window (e.g., 5×5 pixels) to extract the features of the image. In other words, the CNN slides the same window (with the same parameters) to different regions of the image so that we will have the *translation invariance* property. This is important for image processing: in hand-written digit recognition, for example, CNN can well detect the patterns of a digit no matter where the digit is. Technically, such sliding windows are called *convolutional kernels*, whose weights are a part of model parameters and learned during training. This process is shown in Fig. 3.1a.

Another example would be the CNN for natural language processing (Fig. 3.1b). We view a sentence as a sequence of signals in the time axis, word embeddings being the features of different channels. Thus, we can design a neural convolutional kernel (say, spanning three words) and apply it to different positions of a sentence. In this way, we could also capture translation invariant features in natural language sentences. For example, the phrase *"like this movie"* has the same meaning no matter if it appears in the phrase *"do not like this movie."*

We now consider a tree-based convolutional neural network (TBCNN). As indicated by its name, TBCNN applies convolutional kernel over a tree structure. The input is the vector representations of nodes in the tree, where nodes form parent-child

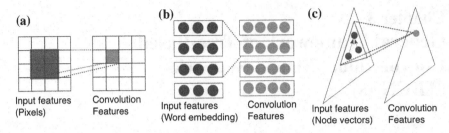

Fig. 3.1 The generic framework of convolutional neural networks (CNNs). **a** CNN for image processing. **b** CNN for image processing. **c** CNN for tree structures

relationship (depicted by dashed arrows in Fig. 3.1c, which are not a part of neural connections).

To design tree-based convolution, we need a fixed-depth sliding window as the convolution kernel, shown as the blue triangle in Fig. 3.1c. The kernel extracts local structural information over the entire tree. Let T be the number of nodes in the kernel, n_e be the dimension of vector representations $x_1, \cdots, x_T$, and n_c be the dimension of convolution.[1] When the kernel is applied to a particular position, tree-based convolution outputs a feature vector y, given by the following generic formula:

$$y = f \left(\sum_{i=1}^{T} W_i x_i + b \right) \tag{3.1}$$

where $W_i \in \mathbb{R}^{n_c \times n_e}$ is the weight matrix for the node x_i, and b is a bias term.

Then the features are aggregated into a fixed-size vector by a pooling operator. A fully-connected hidden layer is also introduced before the final output layer. For classification problems, we use softmax as the output layer.

In this way, the propagation length between input and output layers in TBCNN is fixed and independent of the tree size. Hence, the TBCNN is efficient for structure learning.

It should be noted that, 2D image convolution and 1D (flat) convolution on sentences can be thought of as a special case of Eq. (3.1). For a 2D image, the number of nodes T in Eq. (3.1) is the number of pixels in the kernel ($T = 4$ in Fig. 3.1a). x is the features of a pixel, and for RGB representation, x is a 3 dimensional vector, where each dimension represents the intensity of the pixel in RGB channels. For natural language processing, T is the value of n as in the n-gram model ($T = 3$ in Fig. 3.1b), and x is a word embedding.

---

[1] n_c could also also be thought of as the number of convolution kernels, each kernel outputing one dimensional feature.

3.2 Applications of TBCNN

In this book, we will apply TBCNN to three different types of trees:

- Abstract syntax trees of programs,
- Constituency parse trees of natural language sentences, and
- Dependency parse trees of natural language sentences.

When selecting applications scenarios, we have considered several aspects. First, we should anticipate TBCNN could be applied to different domains, and hence, we chose programming language processing and natural language processing as our application domains. Second, we would hope that, even in one domain, TBCNN could be applied to different variants of trees. Therefore, we applied TBCNN to both constituency trees and dependency trees in NLP.

3.3 Difficulties in Designing TBCNN

To apply TBCNN in different types of trees, there could still be several technical difficulties:

- *How to represent nodes as vectors x?* Different data may have different vector representations. In NLP, for example, each word is represented as an embedding (a continuous, real-valued vector). Such word embeddings could be used for node representations in dependency parse trees, because there is one-to-one correspondence between the words in a sentence and the nodes in the dependency tree. However, in constituency trees, such embeddings are not enough for constituency trees, whose leaf nodes and intermediate nodes have different meanings. A leaf node in a constituency tree corresponds to a word, whereas an intermediate node corresponds to an abstract component of a sentence (e.g, a noun phrase, a verb phrase) which does not have general embeddings.
- *How to determine the number of weights W?* For some trees like a constituency tree, we can always "binarize" it, thus obtaining a binary tree. In such cases, a parent node always has two child nodes, and hence the number of nodes in a convolution window is fixed. However, binarizing abstract syntax trees or dependency trees is not a reasonable approach, and we have to design more elegant ways of allocating weight W. In any case, however, the number of neural parameters should be fixed.
- *How to design pooling?* The output of tree-based convolution is a set of feature vectors, among which exist internal structures as input. Therefore, the size and shape of convolution features vary among different data samples. We need to design pooling mechanisms that can aggregate these convolution features into a fixed-sized vector. A direct approach is "global pooling" that compresses all features into one vector. However, this approach could lose information. Another possible solution is to pool convolution features into several vectors by heuristics. The design criterion of pooling heuristics include

- Features pooled to one vector should be "neighboring" from some perspective, as random pooling does not make much sense.
- Different pools should, in expectation, have a similar number of features. This ensures that the features of different regions in a tree are divided evenly into several pools, which would, intuitively, help to preserve more information.

Chapters 4, 5 and 6 will discuss these technical difficulties in detail when TBCNN is applied to different scenarios.

Chapter 4
TBCNN for Programs' Abstract Syntax Trees

Abstract In this chapter, we will apply the tree-based convolutional neural network (TBCNN) to the source code of programming languages, which we call *programming language processing*. In fact, programming language processing is a hot research topic in the field of software engineering; it has also aroused growing interest in the artificial intelligence community. A distinct characteristic of a program is that it contains rich, explicit, and complicated structural information, necessitating more intensive modeling of structures. In this chapter, we propose a TBCNN variant for programming language processing, where a convolution kernel is designed for programs' abstract syntax trees. We show the effectiveness of TBCNN in two different program analysis tasks: classifying programs according to functionality, and detecting code snippets of certain patterns. TBCNN outperforms baseline methods, including several neural models for NLP.

Keywords Tree-based convolution · Representation learning
Programming language processing · Program analysis

4.1 Introduction

Researchers from various communities are showing growing interest in applying artificial intelligence (AI) techniques to solve software engineering (SE) problems [3, 8, 10]. In the area of SE, analyzing program source code—called *programming language processing* in this paper—is of particular importance.

Although computers can run programs, they do not truly "understand" programs. Analyzing source code provides a way of estimating programs' behavior, functionality, complexity, etc. For instance, automatically detecting source code snippets of certain patterns helps programmers to discover buggy or inefficient algorithms so as to improve code quality. Another example is managing large software repositories,

The contents of this chapter were published in [16]. Copyright©2016, Association for the Advancement of Artificial Intelligence (https://www.aaai.org). Implementation code and the collected dataset are available through our website (https://sites.google.com/site/treebasedcnn/).

© The Author(s) 2018
L. Mou and Z. Jin, *Tree-Based Convolutional Neural Networks*,
SpringerBriefs in Computer Science, https://doi.org/10.1007/978-981-13-1870-2_4

where automatic source code classification and tagging are crucial to software reuse. Programming language processing, in fact, serves as a foundation for many SE tasks, e.g., requirement analysis [9], software development and maintenance [3].

Hindle et al. [12] demonstrate that programming languages, similar to natural languages, also contain abundant statistical properties, which are important for program analysis. These properties are difficult to capture by humans, but justify learning-based approaches for programming language processing. However, existing machine learning program analysis depends largely on feature engineering, which is labor-intensive and *ad hoc* to a specific task, e.g., code clone detection [4], and bug detection [25]. Further, evidence in the machine learning literature suggests that human-engineered features may fail to capture the nature of data, so they may be even worse than automatically learned ones.

The deep neural network, also known as *deep learning*, is a highly automated learning machine. By exploring multiple layers of nonlinear transformation, the deep architecture can automatically learn complicated underlying features, which are crucial to the task of interest. Over the past few years, deep learning has made significant breakthroughs in various fields, such as speech recognition [7], computer vision [14], and natural language processing [5].

Despite some similarities between natural languages and programming languages, there are also obvious differences [18]. Based on a formal language, programs contain rich and explicit structural information. Even though structures also exist in natural languages, they are not as stringent as in programs.

As explained in Sect. 2.3, the natural language sentence "*The dog the stick the fire burned beat bit the cat*" [20] is hard to understand. It complies with all grammar rules, but too many attributive clauses are nested. Hence, it can hardly be understood by people due to the limitation of human intuition capacity. On the contrary, three nested loops are common in programs. The parse tree of a program, in fact, is typically much larger than that of a natural language sentence—there are approximately 190 nodes on average in our experiment, whereas a sentence comprises only 20 words in a sentiment analysis dataset [24]. Further, the grammar rules "alias" neighboring relationships among program components. The statements inside and outside a loop, for example, do not form one semantic group, and thus are not semantically neighboring. On the above basis, we think more effective neural models are in need to capture structural information in programs.

In this chapter, we propose a novel *Tree-Based Convolutional Neural Network* (TBCNN) based on programs' abstract syntax trees (ASTs). We also introduce the notion of "continuous binary trees" and apply dynamic pooling to cope with ASTs of different sizes and shapes. The TBCNN model is a generic architecture, and is applied to two SE tasks in our experiments—classifying programs by functionalities and detecting code snippets of certain patterns. It outperforms baseline methods in both tasks, including the recursive neural network [23] proposed for NLP. To the best of our knowledge, this work is also the first to apply deep neural networks to the field of programming language processing.

4.2 The Proposed Approach

4.2.1 Overview

Programming languages have a natural tree representation—the abstract syntax tree (AST). Figure 4.1 shows the AST of the code snippet "int a = b + 3;".[1] Each node in the AST is an abstract component in program source code. A node p with children $c_1, \ldots, c_n$ represents the constructing process of the component $p \rightarrow c_1 \cdots c_n$.

Figure 4.2 shows the overall architecture of TBCNN. In our model, an AST node is first represented as a distributed, real-valued vector so that the (anonymous) features of the symbols are captured. The vector representations are learned by a coding criterion in one of our papers [19].

We design a set of subtree feature detectors, called the *tree-based convolution kernel*, sliding over the entire AST to extract structural information of a program. Then we apply dynamic pooling to gather information over different parts of the tree. Finally, a hidden layer and an output layer are added. For supervised classification tasks, the activation function of the output layer is softmax.

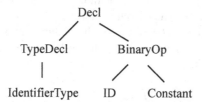

Fig. 4.1 An example of abstract syntax trees (AST). It corresponds to the C code snippet "int a=b+3;" It should be notice that our model takes as input the entire AST of a program, which is typically much larger than a natural language sentence. Reprinted from [16] with permission

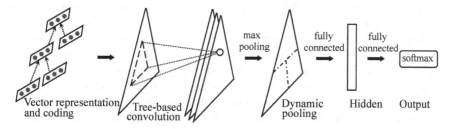

Fig. 4.2 The architecture of the Tree-Based Convolutional Neural Network (TBCNN). The main components in our model include vector representation and coding, tree-based convolution and dynamic pooling; then a fully-connected hidden layer and an output layer (softmax) are added. Reprinted from [16] with permission

---

[1]Parsed by pycparser (https://pypi.python.org/pypi/pycparser/).

In the rest of this section, we first explain the coding criterion for AST nodes' representation learning, serving as a pretraining phase of programming language processing. We then describe the proposed TBCNN model, including a coding layer, a convolutional layer, and a pooling layer. We also provide additional information on dealing with nodes that have varying numbers of child nodes, as in ASTs, by introducing the notion of continuous binary trees.

4.2.2 Representation Learning for AST Nodes

Vector representations, sometimes known as *embeddings*, can capture underlying meanings of discrete symbols, like AST nodes. We propose in [19] an unsupervised method to learn program vector representations by a coding criterion, which serves as a way of pretraining.

A generic criterion for representation learning is "smoothness"—similar symbols have similar feature vectors [2]. For example, the symbols While and For are similar because both of them are related to control flow, particularly, loops. But they are different from ID, since ID probably represents some data. In our scenario, we would like the child nodes' representations to "code" their parent node's via a single neural layer, during which both vector representations and coding weights are learned. Formally, let $\text{vec}(\cdot) \in \mathbb{R}^{N_f}$ be the feature representation of a symbol, where N_f is the feature dimension. For each non-leaf node p and its direct children $c_1, \ldots, c_n$, we would like

$$\text{vec}(p) \approx \tanh \left(\sum_i l_i W_{\text{code},i} \cdot \text{vec}(c_i) + b_{\text{code}} \right) \tag{4.1}$$

where $W_{\text{code},i} \in \mathbb{R}^{N_f \times N_f}$ is the weight matrix corresponding to the node c_i; $b_{\text{code}} \in \mathbb{R}^{N_f}$ is the bias. $l_i = \frac{\#\text{leaves under } c_i}{\#\text{leaves under } p}$ is the coefficient of the weight (weights $W_{\text{code},i}$ are weighted by leaf numbers).

Because different nodes may have different numbers of children, the number of $W_{\text{code},i}$'s is not fixed. To overcome this problem, we introduce the "continuous binary tree," where only two weight matrices W_{code}^l and W_{code}^r serve as model parameters. W_i is a linear combination of the two parameter matrices according to the position of node i. Details are deferred to the last part of this section.

The closeness between $\text{vec}(p)$ and its coded vector is measured by the square of Euclidean distance, i.e.,

$$d = \left\| \text{vec}(p) - \tanh \left(\sum_i l_i W_{\text{code},i} \cdot \text{vec}(c_i) + b_{\text{code}} \right) \right\|_2^2$$

To prevent the pretraining algorithm from learning trivial representations (e.g., all $\mathbf{0}$ vectors will give 0 distance but are meaningless), negative sampling is applied like [6]. For each pretraining data sample $p, c_1, \ldots, c_n$, we substitute one symbol (either p or one of c's) with a random symbol. The distance of the negative sample is denoted as d^-, which should be at least larger than that of the positive training sample plus a margin Δ (set to 1 in our experiment). Thus, the pretraining objective is to

$$\underset{W_{\text{code}}^l, W_{\text{code}}^r, \mathbf{b}_{\text{code}}, \text{vec}(\cdot)}{\text{minimize}} \quad \max\left\{0, \Delta + d - d^-\right\}$$

4.2.3 Coding Layer

Having pretrained the feature vectors for all symbols, we would like to feed them forward to the tree-based convolutional layer for supervised learning. For leaf nodes, they are just the vector representations learned in the pretraining phase. For a non-leaf node p, it has two representations: the one learned in the pretraining phase (left-hand side of Eq. 4.1), and the coded one (right-hand side of Eq. 4.1). They are concatenated and linearly transformed before being fed to the convolutional layer. Let $c_1, \ldots, c_n$ be the children of node p and we denote the combined vector as $\mathbf{p}$. We have

$$\mathbf{p} = W_{\text{comb1}} \cdot \text{vec}(p)$$
$$+ W_{\text{comb2}} \cdot \tanh\left(\sum_i l_i W_{\text{code},i} \cdot \text{vec}(x_i) + \mathbf{b}_{\text{code}}\right)$$

where $W_{\text{comb1}}, W_{\text{comb2}} \in \mathbb{R}^{N_f \times N_f}$ are the parameters for combination. They are initialized as diagonal matrices and then fine-tuned during supervised training.

4.2.4 Tree-Based Convolutional Layer

Now that each symbol in ASTs is represented as a distributed, real-valued vector $\mathbf{x} \in \mathbb{R}^{N_f}$, we apply a set of fixed-depth feature detectors sliding over the entire tree, depicted in Fig. 4.3. The subtree feature detectors can be viewed as convolution with a set of finite support kernels. We call this *tree-based convolution*.

Formally, in a fixed-depth window, if there are n nodes with vector representations $\mathbf{x}_1, \ldots, \mathbf{x}_n$, then the output of the feature detectors is

$$\mathbf{y} = \tanh\left(\sum_{i=1}^n W_{\text{conv},i}\, \mathbf{x}_i + \mathbf{b}_{\text{conv}}\right)$$

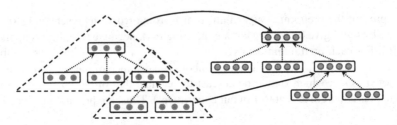

Fig. 4.3 Tree-based convolution. Nodes on the left are the feature vectors of AST nodes. They are either pretrained (for leaf nodes) or combined with pretrained and coded vectors (for non-leaf nodes). Reprinted from [16] with permission

where $y, b_{conv} \in \mathbb{R}^{N_c}$, $W_{conv,i} \in \mathbb{R}^{N_c \times N_f}$ (N_c is the number of feature detectors). **0**'s are padded for nodes at the bottom that do not have as many layers as the feature detectors. In our experiments, the kernel depth is set to 2.

Note that, to deal with varying numbers of children, we also adopt the notion of continuous binary tree. In this scenario, three weight matrices serve as model parameters, namely W^t_{conv}, W^l_{conv}, and W^r_{conv}. $W_{conv,i}$ is a linear combination of these three matrices (explained in detail in the last part of this section).

4.2.5 Dynamic Pooling

After convolution, structural features in an AST are extracted, and a new tree is generated. The new tree has exactly the same shape and size as the original one, which is varying among different programs. Therefore, the extracted features cannot be fed directly to a fixed-size neural layer. Dynamic pooling [21] is applied to deal with this problem.

The simplest approach, perhaps, is to pool all features to one vector. We call this *one-way pooling*. Concretely, the maximum value in each dimension is taken from the features that are detected by tree-based convolution. We also propose an alternative, *three-way pooling*, where features are pooled to 3 parts, TOP, LOWER_LEFT, and LOWER_RIGHT, according to the their positions in the AST

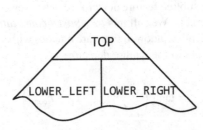

Fig. 4.4 An illustration of 3-way pooling. Reprinted from [16] with permission

(Fig. 4.4). As we shall see from the experimental results, the simple one-way pooling just works as well as three-way pooling. Therefore we adopt one-way pooling in our experiments.

After pooling, the features are fully connected to a hidden layer and then fed to the output layer (softmax) for supervised classification. With the dynamic pooling process, structural features along the entire AST reach the output layer with short paths. Hence, they can be trained effectively by backpropagation.

4.2.6 The "Continuous Binary Tree" Model

As stated, one problem of coding and convolving is that we cannot determine the number of weight matrices because AST nodes have different numbers of children.

One possible solution is the continuous bag-of-words model (CBoW [15], Sect. 2.2.3),[2] but position information will be lost completely. Such approach is also used in [11]. Socher et al. [22] allocate a different weight matrix as parameters for each position; but this method fails to scale up since there will be a huge number of different positions in ASTs.

In our approach, we extend CBoW in that we view any subtree as a "binary" tree, regardless of its size and shape. That is, we have only three weight matrices as parameters for convolution, and two for coding. We call our model a *continuous binary tree*.

Take convolution as an example. The three parameter matrices are W_{conv}^t, W_{conv}^l, and W_{conv}^r (Superscripts t, l, r refer to "top," "left," and "right," respectively). For node x_i in a window, its weight matrix for convolution $W_{conv,i}$ is a linear combination of W_{conv}^t, W_{conv}^l, and W_{conv}^r, with coefficients η_i^t, η_i^l, and η_i^r, respectively. The coefficients are computed according to the relative position of a node in the sliding window. Figure 4.5 is an analogy to the continuous binary tree model. The equations for computing η's are listed as follows.

- $\eta_i^t = \dfrac{d_i - 1}{d - 1}$

  (d_i: the depth of the node i in the sliding window; d: the depth of the window).
- $\eta_i^r = (1 - \eta_i^t)\dfrac{p_i - 1}{n - 1}$

  (p_i: the position of the node i; n: the total number of p's siblings, including p itself).
- $\eta_i^l = (1 - \eta_i^t)(1 - \eta_i^r)$

Likewise, the continuous binary tree for coding has two weight matrices W_{code}^l and W_{code}^r as parameters. The details are not repeated here.

To sum up, the entire parameter set for TBCNN is $\Theta = \{W_{code}^l, W_{code}^r, W_{comb1}, W_{comb2}, W_{conv}^t, W_{conv}^l, W_{conv}^r, W_{hid}, W_{out}, b_{code}, b_{conv}, b_{hid}, b_{out}, \text{vec}(\cdot)\}$, where W_{hid},

---

[2]In their original paper, they do not deal with varying-length data, but their method extends naturally to this scenario. Their method is also mathematically equivalent to average pooling.

Fig. 4.5 An analogy to the continuous binary tree model. In the triangle, the color of a pixel is a combination of three primary colors; in the convolution process, the weight for a node is a combination of three weight parameters, namely W_{conv}^t, W_{conv}^l, and W_{conv}^r. Reprinted from [16] with permission

W_{out}, b_{hid}, and b_{out} are the weights and biases for the hidden and output layers. To set up supervised training, W_{code}^l, W_{code}^r, b_{code}, and vec($\cdot$) are derived from the pretraining phase; W_{comb1} and W_{comb2} are initialized as diagonal matrices; other parameters are initialized randomly.

For classification problems in Sect. 4.3, we apply the cross entropy loss and use stochastic gradient descent with backpropagation.

4.3 Experiments

We first assess the learned vector representations both qualitatively and quantitatively. Then we evaluate TBCNN in two supervised learning tasks, and conduct model analysis.

The dataset of our experiments comes from a pedagogical programming open judge (OJ) system.[3] There are a large number of programming problems on the OJ system. Students submit their source code as the solution to a certain problem; the OJ system automatically judges the validity of submitted source code by running the program. We downloaded the source code and the corresponding programming problems (represented as IDs) as our dataset.

4.3.1 Unsupervised Program Vector Representations

We applied the coding criterion of pretraining to all C code in the OJ system, and obtained AST nodes' vector representations.

---

[3]http://programming.grids.cn. The data are available on our website.

Fig. 4.6 Hierarchical
clustering based on AST
nodes' vector
representations. Reprinted
from [16] with permission

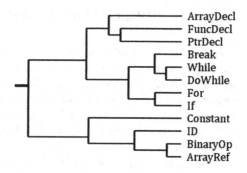

Qualitative analysis. Figure 4.6 illustrates the hierarchical clustering result based on a subset of AST nodes. As demonstrated, the symbols mainly fall into three categories: (1) BinaryOp, ArrayRef, ID, Constant are grouped together since they are related to data reference/manipulation; (2) For, If, While are similar since they are related to control flow; (3) ArrayDecl, FuncDecl, PtrDecl are similar since they are declarations. The result is quite meaningful because it is consistent with human understanding of programs.

Quantitative analysis. We also evaluated pretraining's effect on supervised learning by feeding the learned representations to a program classification task (See next subsection). Figure 4.7 plots the learning curves of both training and validation, which are compared with random initialization. Unsupervised vector representation learning accelerates the supervised training process by nearly 1/3, showing that pretraining does capture underlying features of AST nodes, and that they can emerge high-level features spontaneously during supervised learning. However, pretraining has a limited effect on the final accuracy. One plausible explanation is that the number of AST nodes is small: the pycparser, we used, distinguishes only 44 symbols. Hence, their representations can be adequately tuned in a supervised fashion. Nonetheless, we think the pretraining criterion is effective and beneficial for TBCNN, because training deep neural networks is usually time-consuming, especially when tuning hyperparameters. The pretrained vector representations are used throughout the experiments below.

Fig. 4.7 Learning curves
with and without pretraining.
Reprinted from [16] with
permission

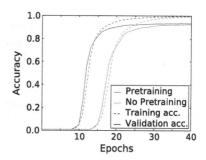

4.3.2 Classifying Programs by Functionalities

Task description. In software engineering, classifying programs by functionalities is an important problem for various software development tasks. For example, in a large software repository (e.g., github, SourceForge), software products are usually organized into categories, a typical criterion for which is by functionalities. With program classification, it becomes feasible to automatically tag a software component newly added into the repository, which is beneficial for software reuse during the development process.

In our experiment, we applied TBCNN to classify source code in the OJ system. The target label of a data sample is one of 104 programming problems (represented as an ID). That is, programs with a same target label have the same functionality. We randomly chose exactly 500 programs in each class, and thus 52,000 samples in total, which were further randomly split by 3:1:1 for training, validation, and testing. Some statistics are shown in Table 4.1.

Hyperparameters. TBCNN's hyperparameters are shown in Table 4.2. Our competing methods include the support vector machine (SVM) and a deep feed-forward neural network based on hand-crafted features, namely bag-of-words (BoW, the counting of each symbol) or bag-of-tree (BoT, the counting of 2-layer subtrees). We also compare our model with the recursive neural network [23]. Hyperparameters for baselines are listed as follows.

Table 4.1 Statistics of our dataset. Reprinted from [16] with permission

Statistics	Mean	Sample std.
# of code lines	36.3	19.0
# of AST nodes	189.6	106.0
Average leaf nodes' depth in an AST	7.6	1.7
Max depth of an AST	12.3	3.2

Table 4.2 TBCNN's hyperparameters. Reprinted from [16] with permission

Hyperparameter	Value	How is the value chosen?
Initial learning rate	0.3	By validation
Learning rate decay	None	Empirically
Embedding dimension	30	Empirically
Convolutional layers' dimension	600	By validation
Penultimate layer's dimension	600	Same as conv layers
l_2 penalty	None	Empirically

- **SVM**. The linear SVM has one hyperparameter C; Radius-basis function (RBF) SVM has two, C and γ. They are tuned by validation over the set $\{\ldots, 1, 0.3, 0.1, 0.03, \ldots\}$ with grid search.
- **DNN**. We applied a 3-layer DNN (excluding input) empirically. The hidden layers' dimension is 300, chosen from $\{100, 300, 1000\}$; learning rates are 0.003 for BoW and 0.03 for BoT, chosen from $\{0.003, \ldots, 0.3\}$ with granularity 3x. ℓ_2 regularization coefficient is 10^{-6} for both BoW and BoT, chosen from $\{10^{-7}, \ldots, 10^{-4}\}$ with granularity 10x, and also no regularization.
- **Recursive neural network**. Recursive units are 600-dimensional, as in our method. The learning rate is chosen from the set $\{\ldots 1.0, 0.3, 0.1 \ldots\}$, and 0.3 yields the highest validation performance.

Results. Table 4.3 presents the results in the 104-label program classification experiment. Using SVM with surface features does distinguish different programs to some extent—for example, a program about string manipulation is different from, say, matrix operation; also, a difficult programming problem necessitates a more complex program, and thus more lines of code and AST nodes. However, their performance is comparatively low.

We tried deep feed-forward neural networks on these features, and achieved accuracies of 76.0–89.7%, comparable to SVMs. Vector averaging with softmax—another neural network-based competing method applied in NLP [13, 24]—yields an accuracy similar to a linear classifier built on BoW features. This is probably because the number of AST symbols is far fewer than words in natural languages, and thus the vector representations (provided non-singular) can be absorbed into the classifier's weights. Comparing these approaches with our method, we deem TBCNN's performance boost is not merely caused by using a better classifier (neural networks versus SVM, say), but also the feature/representation learning nature, which enables automatic structural feature extraction.

Table 4.3 The accuracy of 104-label program classifications. Reprinted from [16] with permission

Group	Method	Test accuracy (%)
Surface features	Linear SVM+BoW	52.0
	RBF SVM+BoW	83.9
	Linear SVM+BoT	72.5
	RBF SVM+BoT	88.2
NN-based approaches	DNN+BoW	76.0
	DNN+BoT	89.7
	Vector average	53.2
	Recursive	84.8
Our method	TBCNN	**94.0**

We also applied the recursive neural network to the program classification task[4]; the recursive net's accuracy is lower than shallow methods (SVM+BoT). Comparing empirical evidence in NLP [23, 24], we observe a degradation of recursive nerual networks' performance if the tree structure is large.

TBCNN outperforms the above methods, yielding an accuracy of 94%. By exploring tree-based convolution, our model is better at capturing programs' structural features, which is important for program analysis.

4.3.3 Detecting Bubble Sort

Task description. To further evaluate our TBCNN model in a more realistic SE scenario, we used it to detect an unhealthy code pattern, bubble sort, which can also be regarded as a (binary) program classification task. Detecting source code of certain patterns is closely related to many SE problems. In this experiment, bubble sort is thought of as unhealthy code because it implements an inefficient algorithm. By identifying such unhealthy code, project managers can refine the implementations during the maintenance process.

Before the experiment, a volunteer annotated, from the OJ system, 109 programs that contain bubble sort, and 109 programs that do not contain bubble sort. They were split 1:1 for validation and testing.

Data augmentation. To train our TBCNN model, a dataset of such scale is insufficient. We propose a simple yet useful data augmentation technique for programs. Concretely, we used the source code of 4k programs in the OJ system as the non-bubble sort class. For each program, we randomly substituted a fragment of program statements with a pre-written bubble sort snippet. Thus we had 8k data samples in total.

Results. We tested our model on the annotated real-world programs. Note that the test samples were written by real-world programmers, and thus the styles and forms of bubble sort snippets may differ from the training set, for example, sorting an integer array versus sorting a user-defined structure, and sorting an array versus sorting two arrays simultaneously. As we see in Table 4.4, bag-of-words features are not illuminating in this classification and yield a low accuracy of 62.3%. Bag-of-trees features are better, and achieve 77.06%. Our model outperforms these methods by more than 10%. This experiment also suggests that neural networks can learn more robust features than just counting surface statistics.

---

[4]We do not use the pretrained vector representations, which are inimical to the recursive neural network: the weight W_{code} encodes children's representation to its candidate parent's; adversely, the high-level nodes in programs (e.g., a function definition) are typically non-informative.

4.3.4 Model Analysis

We now analyze each gadget of TBCNN quantitatively, with the 104-label program classification being our testbed. We report validation accuracies throughout this part.

Effect of coding layer. In the proposed TBCNN model for program analysis, we represent a non-leaf node by combining its coded representation and its pretrained one. We find that, the underneath coding layer can also integrate global information in addition to merely averaging two sources. If we build a tree-based convolutional layer directly on the pretrained vector representations, all structural features are "local," that is, confined in the convolution window. The lack of the integration of global information leads to 2% degradation in performance (See the first and last rows in Table 4.5).

Layers' dimensions. In our experiments, AST nodes' vector representations are set to be 30-dimensional empirically. We chose this small value because AST nodes have only 44 different symbols. Hence, the dimension needs to be, intuitively, smaller than words' vector representations, e.g., 300 in [17]. The dimension of convolution, i.e., the number of feature detectors, was chosen by validation (Fig. 4.8). We

Table 4.4 Accuracy of detecting bubble sort (in percentage). Reprinted from [16] with permission

Classifier	Features	Accuracy
Rand/majority	–	50.0
RBF SVM	Bag-of-words	62.3
RBF SVM	Bag-of-trees	77.1
TBCNN	Learned	**89.1**

Table 4.5 Effect of coding, pooling, and the continuous binary tree. Reprinted from [16] with permission

Model variant	Validation acc.
Coding layer → None	92.3
1-way pooling → 3-way	94.3
Continuous binary tree → CBoW	93.1
TBCNN with the best gadgets	94.4

Fig. 4.8 Validation accuracy versus the number of convolution units. Reprinted from [16] with permission

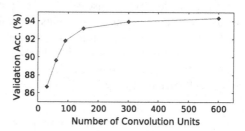

tried several configurations, among which 600-dimensional convolution results in the highest validation accuracy. This analysis also verifies that programs have rich structural information, even though the number of AST symbols is not large. As the rich semantics are emerged by different combinations of AST symbols, we are in need of more feature detectors, that is, a larger convolutional layer.

Effect of pooling layer. We tried two pooling methods in our TBCNN model, and compare them in Table 4.5 (the second and last rows). 3-way pooling is proposed in the hope of preserving features from different parts of the tree. However, as indicated by the experimental result, the simple 1-way pooling works just as fine (even 0.1% higher on the validation set). This suggests that TBCNN is not sensitive to pooling methods, which mainly serve as a necessity for packing varying sized and shaped data. Further development can be addressed in future work.

Effect of continuous binary tree. The continuous binary tree is introduced to treat nodes with different numbers of children, as well as to capture order information of child nodes. We also implemented the continuous bag-of-words (CBoW) model, where child nodes' representations are averaged before convolution. Rows 4 and 5 in Table 4.5 compare our proposed continuous binary tree and the above alternative. The result shows a boost of 1.3% if we consider the order information of child nodes.

4.4   Summary and Discussion

In this chapter, we apply deep neural networks to the field of programming language processing. Due to the rich and explicit tree structures of programs, we propose the novel Tree-Based Convolutional Neural Network (TBCNN) based on abstract syntax trees. In our model, program vector representations are learned by the coding criterion; structural features are detected by the convolutional layer; the continuous binary tree and dynamic pooling enable our model to cope with trees of varying sizes and shapes. Experimental results show the superiority of our model to baseline methods.

We discuss the following issues.

Effect of vector representations. The origin of this work dated back to early 2014, and the papers was preprinted on arXiv in September 2014.[5] Our work was highly influenced by the pretraining-training trend in early days: when we train a neural network, we would like to adopt some unsupervised algorithms to pretrain the weights. However, as discussed before, pretraining is not a necessary step. This is also confirmed by our experiments: pretraining does not improve the ultimate performance. Fortunately, pretraining accelerates the training process and does not have adverse effect in our scenario. Thus we also include the pretraining part in our work. In future work, we need to rethink the effect of pretraining, and it should be only used when necessary.

---

[5]History versions can be found at https://arxiv.org/pdf/1409.3348v1 and https://arxiv.org/pdf/1409.5718v1.

Granularity of program analysis. In this chapter, we analyze the nodes of the abstract syntax tree. This is not a complete representation of programs. For example, the snippets "int a = b + 3;" and "int b = a + 3;" are indistinguishable in the AST representations, as we only care about node types. In fact, we could capture different granularity of programs.

Vector representations map a symbol to a real-valued vector. Possible granularities of the symbol include the character level, token level, etc., discussed below.

- **Character level**. The character-level granularity treats each character as a symbol. Although some research explores character-level modeling for natural language, it is more tricky for programming languages. For example, the token `double` in a C code refers to a data type. But if one writes `doubles`, it is an identifier (e.g., a function name).
- **Token level**. In this level, we learn the representations of all tokens, including types, identifiers, etc. Since programmers can declare their own identifiers in their source code, e.g., `func1`, `func2`, many of the identifiers may appear only a few times, resulting in the undesired data sparseness. Hence, it is improper for direct representation learning at this level.
- **Nodes in ASTs**. In our paper, we learn the representations for nodes in ASTs, e.g., `FuncDef`, `ID`, `Constant`. The AST is more compressed compared with token-level representations. Moreover, we do not distinguish the identifiers, numbers, etc. Therefore, there are only finite many types of nodes in ASTs. The tree structural nature of ASTs also provides opportunities to capture structural information of programs. This level is also used in traditional program analysis like code clone detection [1], vulnerability extrapolation [26], etc.
- **Statement-level, function-level or higher**. Theoretically, a statement, a function or even a program can also be mapped to a real-valued vector, but such representations cannot be trained directly. A possible approach of modeling such complex stuff is by composition. Such research in NLP is often referred to as *compositional semantics* [22]. However, it is very hard to capture the precise semantics; the "semantic barrier" is still not overcome.

The reason why we focus on AST nodes is not because program identifiers are unimportant for program analysis, but we choose AST nodes as a convenient starting point, as we are an early work of deep learning-based program analysis. Moreover, we could focus more on the structural information when modeling programs with TBCNN, ruling out other factors.

References

1. Baxter, I., Yahin, A., Moura, L., Sant'Anna, M., Bier, L.: Clone detection using abstract syntax trees. In: Proceedings of the International Conference on Software Maintenance, pp. 368–377 (1998)
2. Bengio, Y., Courville, A., Vincent, P.: Representation learning: A review and new perspectives. IEEE Trans. Pattern Anal. Mach. Intell. **35**(8), 1798–1828 (2013)

3. Bettenburg, N., Begel, A.: Deciphering the story of software development through frequent pattern mining. In: Proceedings of the 35th International Conference on Software Engineering, pp. 1197–1200 (2013)
4. Chilowicz, M., Duris, E., Roussel, G.: Syntax tree fingerprinting for source code similarity detection. In: Proceedings of the IEEE International Conference on Program Comprehension, pp. 243–247 (2009)
5. Collobert, R., Weston, J.: A unified architecture for natural language processing: Deep neural networks with multitask learning. In: Proceedings of 25th International Conference on Machine Learning, pp. 160–167 (2008)
6. Collobert, R., Weston, J., Bottou, L., Karlen, M., Kavukcuoglu, K., Kuksa, P.: Natural language processing (almost) from scratch. J. Mach. Learn. Res. **12**, 2493–2537 (2011)
7. Dahl, G., Mohamed, A., Hinton, G.: Phone recognition with the mean-covariance restricted Boltzmann machine. In: Advances in Neural Information Processing Systems, pp. 469–477 (2010)
8. Dietz, L., Dallmeier, V., Zeller, A., Scheffer, T.: Localizing bugs in program executions with graphical models. In: Advances in Neural Information Processing Systems, pp. 468–476 (2009)
9. Ghabi, A., Egyed, A.: Code patterns for automatically validating requirements-to-code traces. In: Proceedings of the 27th IEEE/ACM International Conference on Automated Software Engineering, pp. 200–209 (2012)
10. Hao, D., Lan, T., Zhang, H., Guo, C., Zhang, L.: Is this a bug or an obsolete test? In: Proceedings of the European Conference on Object-Oriented Programming, pp. 602–628 (2013)
11. Hermann, K.M., Blunsom, P.: Multilingual models for compositional distributed semantics. In: Proceedings of the 52nd Annual Meeting of the Association for Computational Linguistics, pp. 58–68 (2014)
12. Hindle, A., Barr, E.T., Su, Z., Gabel, M., Devanbu, P.: On the naturalness of software. In: Proceedings of the 34th International Conference on Software Engineering, pp. 837–847 (2012)
13. Kalchbrenner, N., Grefenstette, E., Blunsom, P.: A convolutional neural network for modelling sentences. In: Proceedings of the 52nd Annual Meeting of the Association for Computational Linguistics, pp. 655–665 (2014)
14. Krizhevsky, A., Sutskever, I., Hinton, G.E.: ImageNet classification with deep convolutional neural networks. In: Advances in Neural Information Processing Systems, pp. 1097–1105 (2012)
15. Mikolov, T., Sutskever, I., Chen, K., Corrado, G.S., Dean, J.: Distributed representations of words and phrases and their compositionality. In: Advances in Neural Information Processing Systems, pp. 3111–3119 (2013)
16. Mou, L., Li, G., Zhang, L., Wang, T., Jin, Z.: Convolutional neural networks over tree structures for programming language processing. In: Proceedings of the Thirtieth AAAI Conference on Artificial Intelligence, pp. 1287–1293 (2016)
17. Mou, L., Peng, H., Li, G., Xu, Y., Zhang, L., Jin, Z.: Discriminative neural sentence modeling by tree-based convolution. In: Proceedings of the 2015 Conference on Empirical Methods in Natural Language Processing, pp. 2315–2325 (2015)
18. Pane, J., Ratanamahatana, C., Myers, B.: Studying the language and structure in non-programmers' solutions to programming problems. Int. J. Hum. Comput. Stud. **54**(2), 237–264 (2001)
19. Peng, H., Mou, L., Li, G., Liu, Y., Zhang, L., Jin, Z.: Building program vector representations for deep learning. In: Proceedings of the 8th International Conference on Knowledge Science, Engineering and Management, pp. 547–553 (2015)
20. Pinker, S.: The Language Instinct: The New Science of Language and Mind. Pengiun Press (1994)
21. Socher, R., Huang, E., Pennin, J., Manning, C., Ng, A.: Dynamic pooling and unfolding recursive autoencoders for paraphrase detection. In: Advances in Neural Information Processing Systems, pp. 801–809 (2011)
22. Socher, R., Karpathy, A., Le, Q., Manning, C., Ng, A.Y.: Grounded compositional semantics for finding and describing images with sentences. Trans. Assoc. Comput. Linguist. **2**, 207–218 (2014)

23. Socher, R., Pennington, J., Huang, E., Ng, A., Manning, C.: Semi-supervised recursive autoencoders for predicting sentiment distributions. In: Proceedings of the Conference on Empirical Methods in Natural Language Processing, pp. 151–161 (2011)
24. Socher, R., Perelygin, A., Wu, J., Chuang, J., Manning, C., Ng, A., Potts, C.: Recursive deep models for semantic compositionality over a sentiment treebank. In: Proceedings of the Conference on Empirical Methods in Natural Language Processing, pp. 1631–1642 (2013)
25. Steidl, D., Gode, N.: Feature-based detection of bugs in clones. In: Proceedings of the 7th International Workshop on Software Clones, pp. 76–82 (2013)
26. Yamaguchi, F., Lottmann, M., Rieck, K.: Generalized vulnerability extrapolation using abstract syntax trees. In: Proceedings of 28th Annual Computer Security Applications Conference, pp. 359–368 (2012)

Chapter 5
TBCNN for Constituency Trees in Natural Language Processing

Abstract In this and the following chapters, we will apply the tree-based convolutional neural network (TBCNN) to the natural language processing. This chapter deals with constituency trees of natural language sentences, whereas the next chapter deals with dependency trees. In this chapter, we propose a constituency tree-based convolutional network (c-TBCNN). As usual, c-TBCNN can effectively extract structural information of constituency trees, which is aggregated in one or a few vectors for further information processing. c-TBCNN is applied in two sentence classification tasks: sentiment analysis and question classification. In both experiments, we achieve high performance similar to state-of-the-art models.

Keywords Tree-based convolution · Constituency parsing · Sentence modeling

5.1 Background of Sentence Modeling and Constituency Trees

Discriminative sentence modeling aims to capture the meaning of natural language sentences, and classify sentences into several predefined categories. Let x represent a sentence and y be its label. We would like to build a discriminative model $p(y|x)$. Although some researchers call it *sentence modeling* [7, 19], in machine learning literature, "modeling" is often said in terms of the joint probability $p(y, x)$, for example, language modeling [2]. Therefore, we adopt the terminology *discriminative sentence modeling* to be distinguished from a joint probabilistic modeling.

In fact, the research topic of (discriminative) sentence modeling arises after the prosperity of neural networks. In early years, the domain of NLP is separated by tasks, e.g., topic classification, sentiment analysis, and question classification. Research for a task differs significantly from that for another, although both of them may belong to what we know *feature engineering* today. For example, topic classification can be largely solved by word frequency-inverse document frequency (tf·idf) features [1],

Parts of the contents of this chapter were published in [12]. Copyright © 2015, Association for Computational Linguistics. Implementation code is available through our website (https://sites.google.com/site/tbcnnsentence/).

© The Author(s) 2018
L. Mou and Z. Jin, *Tree-Based Convolutional Neural Networks*,
SpringerBriefs in Computer Science, https://doi.org/10.1007/978-981-13-1870-2_5

or it can be done with unsupervised methods like topic modeling [4]. Sentiment classification sometimes requires sentiment lexicon [4], whereas question classification focuses more on wh-words (e.g., what, when, why, how). To achieve high performance, even more complicated engineering is need as in [14].

Besides feature engineering, kernel methods (like SVMs) are also used for classification problems [13, 18]. In these approaches, We do not define features explicitly, and instead, a *kernel function* specifies an inner product (which can be intuitively thought of as similarity) between two samples. In kernel methods, all properties of data are defined by the kernel, and hence, it is crucial to design the "right" kernel.

Recently, neural networks have brought new opportunities for NLP, showing great potentials in various tasks. For example, word embeddings map discrete features to continuous semantic space, and the relationship between two words can be sometimes represented as vector offset, e.g., "man"−"woman" ≈ "king"−"queen" [11]. Compared with traditional feature engineering and kernel methods, neural networks can be viewed as automatic feature learning or kernel learning.

As said in Chap. 2, several popular neural networks for discriminative sentence modeling include convolutional neural networks (CNNs), recurrent neural networks, and recursive neural networks. CNNs can effectively capture neighboring information and have short propagation paths; however, internal structure information of sentences will be lost. Recursive neural networks can capture structural information to a certain extent, but due to long propagation paths, they suffer from the problem of gradient exploding or vanishing, which results in the difficulty of training [3].[1]

We tackle the question whether we can combine the merits of convolutional and recursive neural networks in the NLP domain. As usual, we would like our model to have short propagation paths as CNNs, but be able to explicitly leverage the parse trees of sentences.

In particular, this chapter deals with constituency trees, where each leaf node corresponds to a word in a sentence and intermediate nodes are abstract components of the sentence (e.g., noun phrase and verb phrase). The definition of constituency trees will be given in Sect. 5.2.1. To represent intermediate nodes in the tree, we propose to nest a recursive neural network inside TBCNN (Sect. 5.2.2). Then we discuss the constituency tree-based convolution and pooling in Sects. 5.2.3 and 5.2.4, respectively. We show experimental results of sentiment analysis and question classification in Sect. 5.3.

---

[1] A recurrent neural network can be viewed as a special case of the recursive neural network, whose structure is a right-most tree.

Fig. 5.1 An example of constituency parse trees. It corresponds to the sentence *"I like the network structure"*

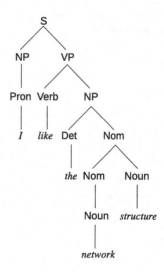

5.2 Proposed Model

5.2.1 Constituency Trees as Input

Figure 5.1 shows an example of the constituency tree corresponding to the sentence, *"I like the network structure."*[2]

In the graph, each leaf node is a word in the sentence, namely, *"I," "like," "the"*... By traversing the tree, we can recover the original sentence. Non-leaf nodes (including intermediate nodes and the root node) represent an abstract component. Table 5.1 illustrates the components in Fig. 5.1 and their representations.

If a parent node p has two child nodes c_1 and c_2, it corresponds to a generation process $p \rightarrow c_1 c_2$. Table 5.2 shows the grammar rules involved in the previous example. It should be noted that constituency trees can be binarized, i.e., a parent node has at most two child nodes, which will help the design of neural network. In our experiment, we used the Stanford parser,[3] and in particular, the binary constituency parser.

5.2.2 Recursively Representing Intermediate Nodes

The first component of our model is a recursive neural newtork [15] that recursively represents non-leaf nodes as vectors. In constituency trees, we do not have pretrained node embeddings because of several reasons:

---

[2]The example is adapted from [6].

[3]http://nlp.stanford.edu/software/lex-parser.shtml.

Table 5.1 Explanation of nodes in Fig. 5.1

Node	Meaning
S	Sentence
NP	Noun phrase
Nom	Nominal
Noun	Noun
VP	Verp phrase
Verb	Verb
Pron	Pronoun

Table 5.2 Grammar rules in Fig. 5.1. The + symbol separates the left and right child nodes

Grammar rule	Example in Fig. 5.1
S → NP VP	*I + like the network structure*
NP → Pron	*I*
NP → Det Nom	*the + network structure*
Nom → Nom Noun	*network + structure*
Nom → Noun	*network*
VP → Verb NP	*like + the network structure*

- There is no pretrained embedding learning algorithm for the nodes in constituency trees. Although we could use the representation learning algorithm introduced in Sect. 4.2.2, the learned embeddings are not in the same space as word embeddings, thus not suitable to a convolutional neural network.
- For an intermediate node in the constituency tree, we care about not only the node type (e.g., Noun), but also its content. In Fig. 5.1, for example, both "*network*" and "*structure*" are a noun, but they are different in meaning. When interacted in tree-based convolution, these two nodes should have different vector representations.

To this end, we propose to nest a recursive neural network underneath c-TBCNN, so that the intermediate nodes can be represented in the same vector space as leaf nodes.

Formally, let c_1 and c_2 be the vector representations of the two child nodes. For leaf nodes, they are pretrained word embeddings. Then the parent node p is represented as

$$p = \tanh(W_1 c_1 + W_2 c_2 + b) \qquad (5.1)$$

where W_1 and W_2 are parameter matrices and b is the bias term. Since the semantic of vector representation is defined by its usage, we can reasonably expect that the encoded representations of intermediate nodes and leaf nodes are in the same semantic space, as they are operated by the same parameters W_1 and W_2 in a recursive fashion.

Fig. 5.2 Constituency
tree-based convolution.
Reprinted from [12] with
permission

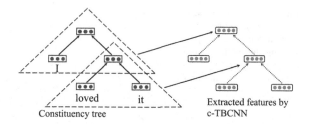

loved   it

Constituency tree

Extracted features by
c-TBCNN

As we focus on tree-based convolution, we pretrain the recursive neural network
as in [15] and then fix its parameters.

5.2.3 Constituency Tree-Based Convolutional Layer

In this part, we design the tree-based convolution for constituency parse trees
(Fig. 5.2). The model variant is denoted as c-TBCNN. For simplicity, we only consider a two-layer subtree. That is to say, the tree-based convolution is defined on a
parent node p and its direct child node c_1 and c_2. Their vector representations, as
defined in Sect. 5.2.2, are denoted as bold letters p, c_l, and c_r, respectively.

Therefore, the general convolution formula (3.1) defined in Chap. 3 becomes

$$y = f\left(W_p^{(c)}p + W_l^{(c)}c_l + W_r^{(c)}c_r + b^{(c)}\right) \tag{5.2}$$

where $W_p^{(c)}$, $W_l^{(c)}$, and $W_r^{(c)}$ are the weight parameters for the parent node and child
nodes, respectively. The superscript (c) implies the constituency tree. There are two
special cases that we need to deal with:

1. A leaf node does not have children, so we feed a zero vector 0 to both c_l and c_r.
2. For some nodes (e.g., NP with a grammar rule NP → Pron), they only have one
   child node. We feed both c_l and c_r with the child node's vector representation.
   This is a special case of the continuous binary tree, defined in Sect. 4.2.6.

As we have adopted the binarized variant of constituency trees, the number of
nodes in a convolution kernel is fixed. Thus, the second technical difficulty in Sect. 3.3
has a trivial solution: assigning weights according to the position in the window.

Constituency tree-based convolution could be extended to an arbitrary depth of
kernels. The computational complexity grows exponentially with respect to the depth,
but is linear to the number of nodes in the kernel. Therefore, given the same amount
of information to process, tree-based convolution, although less flexible, does not
increase the computational complexity, compared with traditional "linear" convolution. In our experiments, we still adopt a 2-layer depth kernel.

Fig. 5.3 The pooling after constituency tree-based convolution: **a** one-way pooling, and **b** three-way pooling. Reprinted from [12] with permission

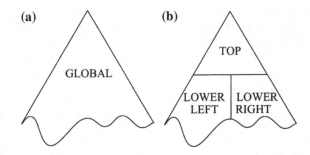

5.2.4 Dynamic Pooling Layer

Similar to Chap. 4, we adopt one-way pooling and three-way pooling for constituency tree-based convolution, as shown in Fig. 5.3.

- One-way pooling, i.e., all features are pooled to one vector (Fig. 5.3a). Here, we also adopt the strategy of max pooling, choosing the maximum value in each dimension.
- Three-way pooling. Suppose the tree as a maximum depth of d. In this method, we pool the top part of the tree (with depth less than $\alpha \cdot d$ to a vector TOP, and the bottom part of the tree to two vectors LOWER_LEFT and LOWER_RIGHT according to the position of the node. (Fig. 5.3b). We are not clear how to pool to more— perhaps an arbitrary number of—vectors while satisfying the criteria introduced in Sect. 3.3. We currently view "three-way pooling" as a fixed mechanism.

In experiments, we will see that three-way pooling is slightly better than one-way pooling, although the improvement is not large. Unless specified, three-way pooling is the default variant for c-TBCNN in our book.

After dynamic pooling, we have a penultimate hidden layer before softmax classification.

5.3 Experiments

We evaluated c-TBCNN on two tasks of discriminative sentence modeling: sentiment analysis (Sect. 5.3.1) and question classification (Sect. 5.3.2). In each experiment, we introduce the task and the dataset that we used; we also provide examples to illustrate the classification task. Then we describe our model settings and present experimental results. The analysis of model variants will be deferred to Chap. 6 for comparison with the dependency tree-based convolution.

Table 5.3 Examples of the sentiment analysis task. Reprinted from [12] with permission

Data sample	Label
Offers that rare combination of entertainment and education	++
An idealistic love story that brings out the latent 15-year-old romantic in everyone	+
Its mysteries are transparently obvious, and it is too slowly paced to be a thriller	−

5.3.1 Sentiment Analysis

5.3.1.1 Task and Dataset

Sentiment analysis is one of the most widely applied tasks of discriminative sentence modeling. In our experiments, we chose the publicly available dataset, Stanford sentiment treebank,[4] which contains more than 100 thousand movie reviews. We followed previous work and adopted the following subtasks:

1. Fine-grained classification. In this setting, the sentiment is classified into five categories: *strongly positive*, *weakly positive*, *neutral*, *weakly negative*, and *strongly negative*.
2. Coarse-grained classification. In this setting, we only consider two labels, (1) *positive*, including both strongly positive and weakly positive reviews; and (2) *negative*, including both strongly negative and weakly negative reviews.

Table 5.3 showcases several illustrative examples of the sentiment analysis task. We also adopted the standard split of the training, validation, and test sets, which contain 8544, 1101, and 2210 sentences, respectively.

Another feature of this dataset is that every valid subsentence (words and sentences) are also manually annotated with sentiment labels. In other words, each node in the constituency tree is labeled with the sentiment corresponding to the phrase under the node, which largely enhances the dataset. A recursive neural network can deal with this case naturally, as we can apply a sentiment classifier recursively on the hidden states. In our TBCNN model, we treat each label (corresponding to a word, a phrase, or an entire sentence) as an independent data sample, following other previous work [7]. In this way, the training set contains more than 150K labels. For validation and testing, we only consider the sentiment of an entire sentence, i.e., the root node in a constituency tree. Our setting is the same as most previous work.

---

[4]http://nlp.stanford.edu/sentiment/.

Table 5.4 Hyperparameters of c-TBCNN for sentiment analysis

Hyperparameter	Value
Dimension of word embeddings	300
Dimension of tree-based convolution	400
Dimension of penultimate hidden layer	200
ℓ_2 penalty for weights	1e-5
ℓ_2 penalty for bias and embeddings	0
Dropout rate for convolution and penultimate hidden layer	50%
Dropout rate for word embeddings and recursive layers	0
Batch size	200
Initial learning rate	3
Learning rate decay	Power decay

5.3.1.2 Experimental Setup

In our experiment, we used ReLU as the activation function, and the pooling method was three-way pooling. The embeddings were pretrained by ourselves using word2vec [11] on the Wikipedia[5] 2014 dump.

Table 5.4 shows other hyperparameters used in our experiments, which were chosen by the validation performance. In particular, we used *dropout* [16] to regularize our model: during training, we randomly set a neuron's activation to 0 by a probability α, whereas during prediction, we compute the activation of all neurons but they are multiplied by α, which is called the *dropout rate*.

5.3.1.3 Results

Table 5.5 shows the main results of c-TBCNN in comparison with previous state-of-the-art methods in the sentiment analysis task. For the fine-grained five-way classification, c-TBCNN achieves 50.4% accuracy. The result is slightly lower than the recursive neural network with long short-term memory interaction reported in [10], but better than other two variants reported in [17, 20]. c-TBCNN is also significantly better than traditional feature-based SVM and naïve Bayes.

For coarse-grained binary classification, our experiment adopted the strategy in [5]: directly transferring the five-way classifier to the binary classification. In other words, we interpret the five-way softmax output as the probability of binary classification.[6] We adopted this setting due to its simplicity in the treatment of the

---

[5]http://en.wikipedia.org.

[6]For the detailed discussion of the binary setting, please refer to http://media.nips.cc/nipsbooks/nipspapers/paper_files/nips27/reviews/521.html.

Table 5.5 Accuracy of sentiment prediction (in percentage). For two-class prediction, "†" indicates that the network is transferred directly from that of five-class. Reprinted from [12] with permission

Group	Method	5-class accuracy	2-class accuracy	Reported in
Baseline	SVM	40.7	79.4	[15]
	Naïve Bayes	41.0	81.8	[15]
CNNs	one-layer convolution	37.4	77.1	[7]
	Deep-CNN	48.5	86.8	[7]
	Non-static	48.0	87.2	[8]
	Multichannel	47.4	**88.1**	[8]
Recursive	Basic	43.2	82.4	[15]
	Matrix-vector	44.4	82.9	[15]
	Tensor	45.7	85.4	[15]
	Tree LSTM (variant 1)	48.9	–	[20]
	Tree LSTM (variant 2)	51.0	88.0	[17]
	Tree LSTM (variant 3)	49.9	88.0	[10]
	Deep RNN	49.8	86.6†	[5]
Recurrent	LSTM	45.8	86.7	[17]
	bi-LSTM	49.1	86.8	[17]
Vector	Word vector avg.	32.7	80.1	[15]
	Paragraph vector	48.7	87.8	[9]
TBCNNs	c-TBCNN	50.4	86.8†	Our

neutral class. In other work, for example [17], they remove the neutral class during prediction; however, they do not mention how they treat the neutral class in training. Our setting is in the disadvantage because the network is not specifically trained for binary classification, but still, it is possible to roughly understand the stability of c-TBCNN. In this setting, we achieve 86.89% accuracy, ranking high in Table 5.5.

In a more controlled setting, we only consider a one-lay of information interaction (linear transformation and nonlinear activation). Our results are consistently better than the recursive neural network (50.4% vs. 43.2% in five-way classification and 86.8% vs. 82.4% in binary classification); also, our results are significantly better than one-layer convolution. These results confirm that the constituency trees are important to sentence modeling, and that c-TBCNN can better capture such structural information than recursive neural networks.

Table 5.6 Examples of the question classification task. Reprinted from [12] with permission

Question	Label
What is the temperature at the center of the earth?	`number`
What state did the Battle of Bighorn take place in?	`location`

Table 5.7 Hyperparameters of c-TBCNN for question classification

Hyperparameter	Value
Dimension of word embeddings	300
Dimension of tree-based convolution	200
Dimension of penultimate hidden layer	100
ℓ_2 penalty for weights	1e-5
ℓ_2 penalty for bias and embeddings	0
Dropout rate for convolution and penultimate hidden layer	50%
Dropout rate for word embeddings and recursive layers	50%
Batch size	25
Initial learning rate	0.5
Learning rate decay	Power decay

5.3.2   Question Classification

We further evaluate the c-TBCNN in a question classification task. The dataset[7] contains 5452 annotated sentences plus 500 test samples in TREC 10. We also use the standard split, like [14]. Target labels contain six classes, namely `abbreviation`, `entity`, `description`, `human`, `location`, and `numeric`. Some examples are shown in Table 5.6.

5.3.2.1   Hyperparameter Settings

The question classification dataset is much smaller than the sentiment analysis one, described in Sect. 5.3.1. By evaluating our model on question classification, we are able to see TBCNN's stability when it is applied to tasks of different dataset sizes.

To prevent overfitting, we adopted a smaller dimension size but a larger dropout rate. Also, we did not fine-tune pretrained word embeddings. Detailed configurations are shown in Table 5.7.

---

[7]Available at http://cogcomp.cs.illinois.edu/Data/QA/QC.

Table 5.8 Accuracy of six-way question classification. Reprinted from [12] with permission

Method	Acc. (%)	Reported in
SVM (10k features + 60 rules)	95.0	[14]
CNN-non-static	93.6	[8]
CNN-mutlichannel	92.2	[8]
RNN	90.2	[19]
Deep-CNN	93.0	[7]
AdaCNN	92.4	[19]
c-TBCNN	94.8	Our implementation

5.3.2.2 Results

Table 5.8 compares c-TBCNN with several existing methods. The first line shows the best previous result, obtained by an SVM with traditional feature engineering [14]. In that work, researchers propose 10 thousand features and 60 hand-crafted rules. On the contrary, c-TBCNN does not make use of any manual rules and features; its architecture is exactly the same as the one for sentiment analysis (Sect. 5.3.1). However, c-TBCNN achieves similar performance to the state-of-the-art result and largely outperforms other neural models.

5.4 Summary and Discussions

In this chapter, we propose a constituency tree-based convolutional neural network (c-TBCNN). Similar to the TBCNN for programs' abstract syntax tree in Chap. 4, c-TBCNN comprises a vector input layer, a convolutional layer, a dynamic pooling layer, a hidden layer, and an output layer.

c-TBCNN is evaluated on two discriminative sentence modeling tasks: sentiment analysis and question classification. Experimental results show that c-TBCNN achieves similar or higher performance, compared with traditional convolutional neural networks, recurrent neural networks, and recursive neural networks.

We discuss the following issue.

Parse Trees of Programs and Natural Language Sentences. The constituency tree of a natural language sentence is similar to the abstract syntax tree of a program in that an intermediate node represents some abstract component in both cases. Therefore, the tree-based convolution is also similar in these two scenarios. We will see in the next chapter that dependency parse trees differ largely from constituency parse trees, and thus TBCNN variants also differ from each other.

However, it should also be mentioned that, in this chapter, we use the binarized constituency tree for natural language sentences, which is not suitable for programs. For example, for the node StatemntList in an abstract syntax tree, it has a list of statements `Statement_1`, `Statement_2`, `Statement_3`, ... If we binarize such trees, we will obtain a left-most or right-most tree. In a program, such statement lists could be long, and thus binarization is not appropriate. In natural language, the child list will not be too long because of the limit of human intuition. Therefore, binarizing constituency trees is a natural treatment, which is also used in the recursive neural network [15].

References

1. Aizawa, A.: An information-theoretic perspective of TF-IDF measures. Inf. Process. Manag. **39**(1), 45–65 (2003)
2. Bengio, Y., Ducharme, R., Vincent, P., Janvin, C.: A neural probabilistic language model. J. Mach. Learn. Res. **3**, 1137–1155 (2003)
3. Erhan, D., Manzagol, P., Bengio, Y., Bengio, S., Vincent, P.: The difficulty of training deep architectures and the effect of unsupervised pre-training. In: Proceedings of International Conference on Artificial Intelligence and Statistics, pp. 153–160 (2009)
4. Hatzivassiloglou, V., McKeown, K.: Predicting the semantic orientation of adjectives. In: Proceedings of the 8th Conference on European Chapter of the Association for Computational Linguistics, pp. 174–181 (1997)
5. Irsoy, O., Cardie, C.: Deep recursive neural networks for compositionality in language. In: Advances in Neural Information Processing Systems, pp. 2096–2104 (2014)
6. Jurafsky, D., Martin, J.: Speech and Language Processing. Pearson Education (2000)
7. Kalchbrenner, N., Grefenstette, E., Blunsom, P.: A convolutional neural network for modelling sentences. In: Proceedings of the 52nd Annual Meeting of the Association for Computational Linguistics, pp. 655–665 (2014)
8. Kim, Y.: Convolutional neural networks for sentence classification. In: Proceedings of the 2014 Conference on Empirical Methods in Natural Language Processing, pp. 1746–1751 (2014)
9. Le, Q., Mikolov, T.: Distributed representations of sentences and documents. In: Proceedings of the International Conference on Machine Learning, pp. 1188–1196 (2014)
10. Le, P., Zuidema, W.: Compositional distributional semantics with long short term memory (2015). arXiv preprint arXiv:1503.02510
11. Mikolov, T., Sutskever, I., Chen, K., Corrado, G., Dean, J.: Distributed representations of words and phrases and their compositionality. In: Advances in Neural Information Processing Systems, pp. 3111–3119 (2013)
12. Mou, L., Peng, H., Li, G., Xu, Y., Zhang, L., Jin, Z.: Discriminative neural sentence modeling by tree-based convolution. In: Proceedings of the 2015 Conference on Empirical Methods in Natural Language Processing, pp. 2315–2325 (2015)
13. Reichartz, F., Korte, H., Paass, G.: Semantic relation extraction with kernels over typed dependency trees. In: Proceedings of the 16th ACM SIGKDD International Conference on Knowledge Discovery and Data Mining, pp. 773–782 (2010)
14. Silva, J., Coheur, L., Mendes, A., Wichert, A.: From symbolic to sub-symbolic information in question classification. Artif. Intell. Rev. **35**(2), 137–154 (2011)
15. Socher, R., Pennington, J., Huang, E., Ng, A., Manning, C.: Semi-supervised recursive autoencoders for predicting sentiment distributions. In: Proceedings of the Conference on Empirical Methods in Natural Language Processing, pp. 151–161 (2011)
16. Srivastava, N., Hinton, G., Krizhevsky, A., Sutskever, I., Salakhutdinov, R.: Dropout: a simple way to prevent neural networks from overfitting. J. Mach. Learn. Res. **15**(1), 1929–1958 (2014)
17. Tai, K.S., Socher, R., Manning, C.D.: Improved semantic representations from tree-structured long short-term memory networks. In: Proceedings of the 53rd Annual Meeting of the Association for Computational Linguistics and the 7th International Joint Conference on Natural Language Processing (Volume 1: Long Papers), pp. 1556–1566 (2015)

18. Zelenko, D., Aone, C., Richardella, A.: Kernel methods for relation extraction. J. Mach. Learn. Res. **3**, 1083–1106 (2003)
19. Zhao, H., Lu, Z., Poupart, P.: Self-adaptive hierarchical sentence model. In: Proceedings of Intentional Joint Conference in Artificial Intelligence, pp. 4069–4076 (2015)
20. Zhu, X., Sobihani, P., Guo, H.: Long short-term memory over tree structures. In: Proceedings of the 32nd International Conference on Machine Learning, pp. 1604–1612 (2015)

Chapter 6
TBCNN for Dependency Trees in Natural Language Processing

Abstract This chapter applies tree-based convolution to the dependency parse trees of natural language sentences, resulting in a new variant d-TBCNN. Since dependency trees are different from abstract syntax trees in Chap. 4 and constituency trees in Chap. 5, we need to design new model gadgets for d-TBCNN. The model is evaluated on two sentence classification tasks (sentiment analysis and question classification) and a sentence matching task. In the sentence classification tasks, d-TBCNN outperforms previous state-of-the-art results, whereas in the sentence matching task, d-TBCNN achieves comparable performance to the previous state-of-the-art model, which has a higher matching complexity.

Keywords Tree-based convolution · Dependency parsing · Sentence modeling
Sentence matching

6.1 Background of Dependency Trees

Dependency parsing is another tree structural representation of natural language sentences. Different from constituency parse trees, every node (including leaves, intermediate nodes, and root nodes) corresponds to a word in the sentence. A parent–child relation indicates that a word is dependent on another. Therefore, dependency parsing is more compact than constituency parsing, with a shallower structure. Also, multiple words can depend on a same word, and hence, the nodes in a dependency tree may have different numbers of child nodes.

To cope with the above problems, we propose a dependency tree-based convolutional neural network, denoted as d-TBCNN (Sect. 6.2). The main difference of d-TBCNN and previous variants is that d-TBCNN indexes convolution weights by the dependency type of a word as opposed to its position, and that during pooling d-TBCNN reconstructs the order of words as in the sentence and perform pooling by an "equal allocation" strategy.

Parts of the contents of this chapter were published in [17, 18]. Copyright © 2015, 2016, Association for Computational Linguistics. Implementation code is available through our websites (https://sites. google.com/site/tbcnnsentence/ and https://sites.google.com/site/tbcnninference/).

© The Author(s) 2018
L. Mou and Z. Jin, *Tree-Based Convolutional Neural Networks*,
SpringerBriefs in Computer Science, https://doi.org/10.1007/978-981-13-1870-2_6

Fig. 6.1 An example of
dependency parse trees
corresponding to the
sentence, "*We propose the
model in the book*"

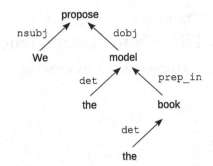

To evaluate the effectiveness of d-TBCNN, we also conducted experiments on discriminative sentence modeling, namely, sentiment analysis and question classification, as introduced in Chap. 5. Experimental results show that the d-TBCNN proposed in this chapter outperforms previous models (Sect. 6.3.1).

To further evaluate our model, we applied it to a sentence matching task that has a massive publicly available dataset, Stanford Natural Language Inference (SNLI) dataset [2]. In this task, we achieve similar results to the state-of-the-art model, but with a lower matching matching complexity. The model and experiments are described in Sects. 6.2.4 and 6.3.2, respectively.

We also conducted in-depth model analysis of both d-TBCNN and c-TBCNN in Sect. 6.3.3, and visualization in Sect. 6.3.4.

6.2 Proposed Model

6.2.1 Dependency Trees as Input

Figure 6.1 shows the dependency tree of the sentence, "*We propose the model in the book*."[1]

In the dependency parse tree, an edge $a \xrightarrow{r} b$ represents the word a being dependent on the word b. Table 6.1 summarizes the dependency types and their meanings in Fig. 6.1.

Intuitively, the relationship $a \rightarrow b$ is defined such that b is the more "central" word between a and b and that the meaning of a depends on the existence of b. Consider the phrase "*in the book*" in the previous example. The word "*the*" is dependent on "*book*." This is because "*the*" is an article that "modifies" the word "*book*." We can reasonably argue that without the word "*book*," "*the*" has no meaning in the sentence, but that even without "*the*," the meaning of "*book*" can stand by its own. Moreover, the *type* of this dependency relation is that "*the*" is a determiner of "*book*." Thus we have *the* $\xrightarrow{det}$ *book*.

---

[1]The example is adapted from [12].

Table 6.1 The dependency types in Fig. 6.1 and their meanings

Dependency type	Meaning
nsubj	Nominal subject
dobj	Direct object
det	Determiner
prep_in	Preposition (*in*)

We have several remarks

- Generally speaking, the verb in the main clause is the root node in the dependency tree, i.e., it does not depend on other words. Although it is less obvious than the example of "*the book*," linguistics believe that the verb plays the most important role, especially more important than the subject and object.
- In some literature, e.g., the textbook by Jurafsky and Martin [12], the arrows are in the opposite directions to our notation. In their notation, $a \rightarrow b$ represents a governing b, i.e., b being dependent on a. In our book, however, we adopt a difficult direction, so that it complies with the tree structure as in the constituency tree in Chap. 5. In fact, this is only a notational difference, and does not affect the nature of dependency parsing.
- The last dependency type in Table 6.1 (prep_in) indicates that the word "*model*" is dependent on "*book*" by the meaning of the preposition "*in*." This is a *collapsed* variant of dependency parsing. In this chapter, we use this variant, and parse all sentences by the Stanford parser.[2]

With the dependency tree as the skeleton, d-TBCNN accepts nodes' vector representations as input. Here, we have a natural solution to the first technical difficulty raised in Sect. 3.3: since each node in the dependency tree corresponds to a word, we simply use pretrained word embeddings as input, which are then fine-tuned by backpropagation.

6.2.2 Convolutional Layer

The nature of dependency representation leads to d-TBCNN's major difference from traditional convolution: there exist nodes with different numbers of child nodes. This causes trouble if we associate weight parameters according to positions in the window, which is standard for traditional convolution [5] or c-TBCNN (Fig. 6.2).

To overcome the problem, we extend the notion of convolution by assigning weights according to dependency types (e.g., nsubj) rather than positions. We believe this strategy makes much sense because dependency types reflect the relationship between a governing word and its child words. To be concrete, the generic convolution formula (3.1) for d-TBCNN becomes

---

[2]http://nlp.stanford.edu/software/lex-parser.shtml.

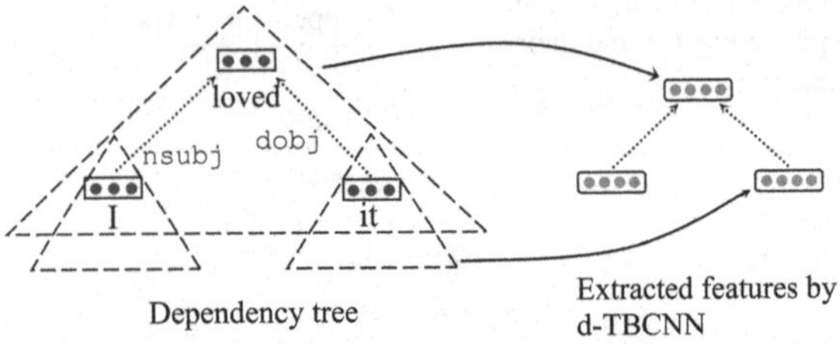

Fig. 6.2 Dependency
tree-based convolution.
Reprinted from [17] with
permission

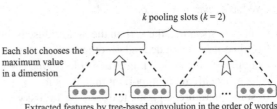

Dependency tree   Extracted features by
                                             d-TBCNN

Fig. 6.3 k-slot pooling for
d-TBCNN. Reprinted
from [17] with permission

k pooling slots ($k = 2$)

Each slot chooses the
maximum value
in a dimension

Extracted features by tree-based convolution in the order of words

$$y = f\left(W_p^{(d)} p + \sum_{i=1}^{n} W_{r[c_i]}^{(d)} c_i + b^{(d)}\right) \qquad (6.1)$$

where $W_p^{(d)}$ is the weight parameter for the parent p (governing word); $W_{r[c_i]}^{(d)}$ is
the weight for child c_i, which has grammatical relationship $r[c_i]$ to its parent, p.
Superscript (d) indicates the parameters are for d-TBCNN. Note that we keep 15
most frequently occurred dependency types; others appearing rarely in the corpus
are mapped to one shared weight matrix.

6.2.3   Dynamic Pooling Layer

In d-TBCNN, we still need a dynamic pooling layer to aggregate extracted features,
as different sentences may have different lengths and tree structures.

Similar to Sects. 4.2.5 and 5.2.4, we could always use the one-way pooling (also
called *global pooling*) that pools all features to a single fixed-size vector. In addition,
this section provides an alternative of k-way pooling for d-TBCNN, as shown in
Fig. 6.3.

Different from constituency trees, nodes in dependency trees are one-one corresponding to words in a sentence. Thus, a total order on features (after convolution)
can be defined according to their corresponding word orders. For k-slot pooling, we
can adopt an "equal allocation" strategy, shown in Fig. 6.3. Let i be the position of
a word in a sentence ($i = 1, 2, \ldots, n$). Its extracted feature vector is pooled to the
j-th slot, if

$$(j-1)\frac{n}{k} \leq i \leq j\frac{n}{k} \qquad (6.2)$$

Table 6.2 Examples of the relations between a premise and a hypothesis: Entailment, Contradiction, and Neutral (irrelevant). Reprinted from [18] with permission

Premise	Two men on bicycles competing in a race	
Hypothesis	People are riding bikes	Entailment
	Men are riding bicycles on the streets	Contradiction
	A few people are catching fish	Neutral

We assess the effectiveness of pooling quantitatively in Sect. 6.3.3. As we shall see by the experimental results, complicated pooling methods do preserve more information along tree structures to some extent, but the effect is not large. TBCNNs are not very sensitive to pooling methods.

6.2.4 Applying d-TBCNN to Sentence Matching

In addition to discriminative sentence modeling, we further apply d-TBCNN to sentence matching. In particular, we consider the natural language inference (NLI) task, which aims to recognizing entailment and contradiction between two sentences (called a *premise* and a *hypothesis*). Provided with a premise sentence, the task is to judge whether the hypothesis can be inferred (entailment), or the hypothesis cannot be true (contradiction). A third class Neutral is also introduced to indicate that two sentences are not related. Several examples are illustrated in Table 6.2.

NLI is important to natural language understanding and has wide applications in NLP, e.g., question-answering [7] and automatic summarization [14]. Moreover, NLI is also related to other tasks of sentence pair modeling, including paraphrase detection [10], relation recognition of discourse units [3].[3]

Traditional approaches to NLI mainly fall into two groups: feature-based methods and formal reasoning methods. Feature-based approaches typically leverage machine learning models, but require intensive human engineering to represent lexical and syntactic information in two sentences [8, 15]. Formal reasoning, on the other hand, converts a sentence into a formal logical representation and uses interpreters to search for a proof. However, such approaches are limited in terms of scope and accuracy [1].

The renewed prosperity of neural networks has made significant achievements in various NLP applications, including discriminative sentence modeling as well as sentence matching. A typical neural architecture to model sentence pairs is the

---

[3]Rigorously, NLI is not aimed to "match" two sentences in the sense of information retrieval. However, NLI is indeed to model a pair of sentences, and can be thought of as matching two sentences in terms of entailment. Therefore, we are happy to abuse the terminology in our book.

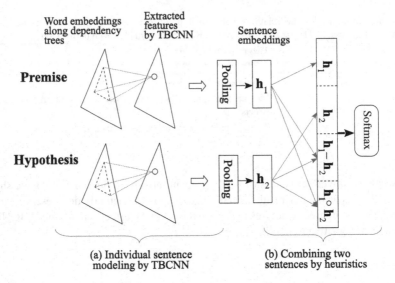

Fig. 6.4 TBCNN applied to natural language inference. Reprinted from [18] with permission

"Siamese" structure [4], which involves an underlying sentence model and a matching layer to determine the relationship between two sentences. In this section, we focus on the question whether TBCNN can be harnessed to model sentence pairs for implicit logical inference, as is in the NLI task.

Figure 6.4 shows the framework of TBCNN applied in the NLI task. Our model comprises the following two components:

1. A sentence model extracts a feature vector h for both premise and hypothesis, shown in Fig. 6.4a. Here, we use dependency tree-based convolution with global pooling as the underlying model, and we tie the parameters for both premise's and hypothesis's TBCNN, as we hope this part to capture general semantics of sentences.
2. A matching layer combines two sentences' information by heuristics (Fig. 6.4b). After individual sentence modeling, we design a sentence matching layer to aggregate information. We use simple heuristics, including concatenation, element-wise product and difference, which are effective and efficient. Since the underlying d-TBCNN has been introduced in detail in previous subsections, this part mainly focuses on the matching heuristics.

Formally, we have three matching heuristics

1. Concatenation of the two sentence vectors,
2. Element-wise product, and
3. Element-wise difference.

The first heuristic follows the most standard procedure of the "Siamese" architectures, while the latter two are certain measures of "similarity" or "closeness." These matching layers are further concatenated (Fig. 6.4b), given by

$$m = \begin{pmatrix} h_1 \\ h_2 \\ h_1 \circ h_2 \\ h_1 - h_2 \end{pmatrix} \tag{6.3}$$

where $h_1 \in \mathbb{R}^{n_c}$ and $h_2 \in \mathbb{R}^{n_c}$ are the sentence vectors of the premise and hypothesis, respectively, and "$\circ$" denotes element-wise product. $m \in \mathbb{R}^{4n_c}$ is the output of the matching layer. Here, n_c denotes the dimension of tree-based convolution.

We would like to point out that, with subsequent linear transformation, element-wise difference is a special case of concatenation. If we assume the subsequent transformation takes the form of $W[h_1 \ h_2]^\top$, where $W = [W_1 \ W_2]$ is the weights for concatenated sentence representations, then element-wise difference can be viewed as such that $W_0(h_1 - h_2) = [W_0 \ -W_0][h_1 \ h_2]^\top$. ($W_0$ is the weights corresponding to element-wise difference.) Thus, our third heuristic can be absorbed into the first one in terms of model capacity. However, as will be shown in the experiment, explicitly specifying this heuristic significantly improves the performance, indicating that optimization differs, despite the same model capacity. Word embedding studies show that linear offset of vectors can capture relationships between two words [16], but it has not been exploited in sentence–pair relation recognition. Although element-wise distance is used to detect paraphrase in [9], it mainly reflects "similarity" information. Our study verifies that vector offset is useful in capturing generic sentence relationships, akin to the word analogy task.

Moreover, since these heuristics operate on the features exacted by d-TBCNN, it further verifies the effectiveness of tree-based convolution in tasks other than discriminative sentence modeling.

6.3 Experiments

In this section, we first present the overall performance of d-TBCNN for discriminative sentence modeling, including results in sentiment analysis and question classification (Sect. 6.3.1); then we show d-TBCNN's performance when applied to sentence matching (Sect. 6.3.2). Section 6.3.3 performs in-depth analysis. For the purposes of comparison, we also include the constituency tree-based convolutional neural network (c-TBCNN, Chap. 5). Section 6.3.4 conducts visualization to better understand the proposed tree-based convolution.

Table 6.3 Hyperparameters of d-TBCNN for sentiment analysis and question classification

Hyperparameter	Value in sentiment analysis	Value in question classification
Dimension of word embeddings	300	300 (fixed)
Dimension of tree-based convolution	300	30
Dimension of penultimate hidden layer	200	25
ℓ_2 penalty for weights	1e-5	0
ℓ_2 penalty for bias and embeddings	0	0
Dropout rate for convolution and penultimate layer	50%	5%
Dropout rate for word embeddings and recursive layers	40%	30%
Batch size	200	25
Initial learning rate	3	0.3
Learning rate decay	Power decay	Power decay

6.3.1 Discriminative Sentence Modeling

In this section, we apply d-TBCNN to two tasks of discriminative sentence modeling, namely, sentiment analysis and question classification. The tasks have been described in Sect. 5.3, and are not repeated here.

Table 6.3 lists the hyperparameters of d-TBCNN in the two tasks. Since the hyperparameters of our models are tuned by validation in each task, they are different from Tables 5.4 and 5.7 for constituency trees. However, they are not the major difference.

Table 6.4 shows the performance of d-TBCNN for sentence classification, compared with best results published in previous papers (described in Sect. 5.3). Experimental results show that d-TBCNN consistently outperforms c-TBCNN in all experiments, achieving new state-of-the-art results in both experiments (except the two-way sentiment classification setting).

In should be emphasized that, to the best of our knowledge, this is the first time that neural networks beat dedicated human engineering in this question classification task. Previous work—based on an SVM classifier with more than 60 features and 10 thousand rules—achieves 95% accuracy. On the contrary, we design the elegant neural structure and outperform previous work by 1% (Table 6.5).

Table 6.4 Performance of TBCNN for sentence classification, compared with state-of-the-art (SOTA) models

Model	Five-way sentiment classification	Two-way sentiment classification	Question classification
SOTA	51.0 [22]	**88.1** [13]	95.0 [20]
c-TBCNN	50.4	86.8	94.8
d-TBCNN	**51.4**	87.9	**96.0**

Table 6.5 Statistics of the Stanford Natural Language Inference dataset where each sentence is parsed into a dependency parse tree. Reprinted from [18] with permission

Statistics	Mean	Std
# nodes	8.59	4.14
Max depth	3.93	1.13
Avg leaf depth	3.13	0.65
Avg node depth	2.60	0.54

6.3.2 Sentence Matching

6.3.2.1 Task and Dataset

To evaluate our TBCNN model in the sentence matching task, we used the newly published Stanford Natural Language Inference (SNLI) dataset [2].[4] The dataset is constructed by crowdsourced efforts, each sentence written by humans. Moreover, the SNLI dataset is larger than previous resources in magnitude, and hence is particularly suitable for comparing neural models. The target labels comprise three classes: Entailment, Contradiction, and Neutral (two irrelevant sentences). We applied the standard train/validation/test split, containing 550k, 10k, and 10k samples, respectively. Figure 6.5 presents additional dataset statistics, especially those relevant to dependency parse trees.[5]

All our neural layers, including embeddings, were set to 300 dimensions. The model is mostly robust when the dimension is large, e.g., several hundred [5]. Word embeddings were pretrained ourselves by word2vec on the English Wikipedia corpus and fine-tuned during training as a part of model parameters. We used stochastic gradient descent with a batch size of 50. Initial learning rate was set to 1, and a power decay was applied. We applied ℓ_2 penalty of 3×10^{-4}; dropout was chosen by validation with a granularity of 0.1, tuned as in Fig. 6.5. We see that a large dropout rate (≥ 0.3) hurts the performance (and also makes training slow) for such a large dataset, different from small datasets in other tasks (e.g., Sect. 6.3.1).

---

[4]http://nlp.stanford.edu/projects/snli/.

[5]We applied *collapsed* dependency trees, where prepositions and conjunctions are annotated on the dependency relations, but these auxiliary words themselves are removed.

Fig. 6.5 Validation accuracy versus dropout rate. Reprinted from [18] with permission

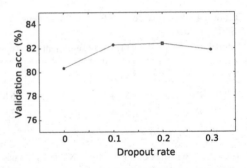

Table 6.6 Accuracy of the TBCNN-pair model in comparison with previous results ([b]Bowman et al. [2]; [v]Vendrov et al. [23]; [r]Rocktäschel et al. [19]). "cat" refers to concatenation;"-" and "○" denote element-wise difference and product, respectively. Reprinted from [18] with permission

Model	Test acc. (%)	Matching complexity
Unlexicalized features[b]	50.4	$\mathcal{O}(1)$
Lexicalized features[b]	78.2	
Vector sum + MLP[b]	75.3	
Vanilla RNN + MLP[b]	72.2	
LSTM RNN + MLP[b]	77.6	
CNN + cat	77.0	
GRU w/ skip-thought pretraining[v]	81.4	
TBCNN-pair + cat	79.3	
TBCNN-pair + cat,○,-	**82.1**	
Single-chain LSTM RNNs[r]	81.4	$\mathcal{O}(n)$
+ static attention[r]	**82.4**	
LSTM + word-by-word attention[r]	**83.5**	$\mathcal{O}(n^2)$

6.3.2.2   Accuracy

Table 6.6 compares our model with previous results. As seen, the TBCNN sentence pair model, followed by simple concatenation alone, outperforms existing sentence encoding-based approaches (without pretraining), including a feature-rich method using six groups of human-engineered features, long short-term memory (LSTM)-based RNNs, and traditional CNNs. This verifies the rationale for using tree-based convolution as the sentence-level neural model for NLI.

Table 6.7 compares different heuristics of matching. We first analyze each heuristic separately: using element-wise product alone is significantly worse than concatenation or element-wise difference; the latter two are comparable to each other.

Combining different matching heuristics improves the result: the TBCNN-pair model with concatenation, element-wise product and difference yields the highest

Table 6.7 Validation and test accuracies of TBCNN-pair variants (in percentage). Reprinted from [18] with permission

Model variant	Valid acc.	Test acc.
TBCNN+o	73.8	72.5
TBCNN+-	79.9	79.3
TBCNN+cat	80.8	79.3
TBCNN+cat,o	81.6	80.7
TBCNN+cat,-	81.7	81.6
TBCNN+cat,o,-	**82.4**	**82.1**

performance of 82.1%. As analyzed in Sect. 6.2.4, the element-wise difference matching layer does not add to model complexity and can be absorbed as a special case into simple concatenation. However, explicitly using such heuristic yields an accuracy boost of 1–2%. Further applying element-wise product improves the accuracy by another 0.5%.

The full TBCNN-pair model outperforms all existing sentence encoding-based approaches, including a 1024d gated recurrent unit (GRU)-based RNN with "skip-thought" pretraining [23]. The results obtained by our model are also comparable to several attention-based LSTMs, which are more computationally intensive than ours in terms of complexity order.

6.3.2.3 Complexity Concerns

For most sentence models including TBCNN, the overall complexity is at least $\mathcal{O}(n)$. However, an efficient matching approach is still important, especially to *retrieval-and-reranking* systems [24]. For example, in a retrieval-based question-answering or conversation system, we can largely reduce response time by performing sentence matching based on precomputed candidates' embeddings. By contrast, context-aware matching approaches [19] involve processing each candidate given a new user-issued query, which is time-consuming in terms of most industrial products.

In our experiments, the matching part (Fig. 6.4b) counts 1.71% of the total time during prediction (C++ implementation on CPU), showing the potential applications of our approach in efficient retrieval of semantically related sentences.

6.3.3 Model Analysis

In this subsection, we have in-depth analysis of TBCNN in natural language processing to show the effect of different variants. Our testbed is five-way sentiment classification, and for the sake of comparison, we include both c-TBCNN in Chap. 5 and d-TBCNN in this chapter.

6.3.3.1   The Effect of Pooling

The extracted features by tree-based convolution have varying topologies in size and shape. We propose in Sects. 5.2.4 and 6.2.3 several heuristics for pooling. This subsection aims to provide a fair comparison among these pooling methods.

One reasonable protocol for comparison is to tune all hyperparameters for each setting and compare the highest accuracy. This methodology, however, is time-consuming, and depends largely on the quality of hyperparameter tuning. An alternative is to predefine a set of meaningful hyperparameters and report the accuracy under the same setting. In this experiment, we chose the latter protocol, where hidden layers are all 300-dimensional; no ℓ_2 penalty is added. Each configuration was run five times with different random initializations. We summarize the mean and standard deviation in Table 6.8.

As the results imply, complicated pooling is better than global pooling to some degree in both model variants. But the effect is not strong; our models are not that sensitive to pooling methods, which mainly serve as a necessity for dealing with varying-structured data. In the results reported before, we applied three-way pooling for c-TBCNN and two-way pooling for d-TBCNN.

Comparing with other studies in the literature, we also notice that pooling is very efficient in information gathering. Irsoy and Cardie [11] report 200 epochs for training a deep RNN, which achieves 49.8% accuracy in the five-class sentiment classification. Our TBCNNs are typically trained within 25 epochs.

6.3.3.2   The Effect of Sentence Lengths

We analyze how sentence lengths affect our models. Sentences are split into seven groups by length, with a granularity of five. A few too long or too short sentences are grouped together for smoothing; the number of sentences in each group varies from 126 to 457. Figure 6.6 presents accuracies versus lengths in TBCNNs. For comparison, we also reimplemented the recursive neural network, achieving 42.7% overall accuracy, slightly worse than 43.2% reported in [21]. Thus, we think our reimplementation is fair and that the comparison is meaningful.

We observe that c-TBCNN and d-TBCNN have very similar behaviors. They consistently outperform the recursive neural network in all scenarios. We also notice that the gap, between TBCNNs and the recursive neural network, increases when sentences contain more than 20 words. This result confirms that for long sentences, the propagation paths in a recursive network are deep, causing its difficulty in information processing. By contrast, our models explore structural information more effectively with tree-based convolution. As information from any part of the tree can

Table 6.8 Accuracies of different pooling methods, averaged over 5 random initializations. We manually chose meaningful hyperparameters in advance to make a fair comparison. This leads to performance degradation (1–2%) vis-a-vis Tables 5.5 and 6.4. Reprinted from [17] with permission

Model	Pooling method	Five-class accuracy (%)
c-TBCNN	Global	48.48 ± 0.54
	3-way	48.69 ± 0.40
d-TBCNN	Global	49.39 ± 0.24
	2-way	49.94 ± 0.63

Fig. 6.6 Accuracies versus sentence lengths. In the legend, RNN refers to the recursive neural network. Reprinted from [17] with permission

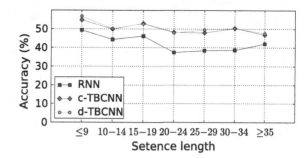

propagate to the output layer with short paths, TBCNNs are more capable for sentence modeling, especially for long sentences.

6.3.4 Visualization

Visualization is important to understanding the mechanism of neural networks. For TBCNNs, we would like to see how the extracted features (after convolution) are further processed by the max pooling layer, and ultimately related to the supervised task.

To show this, we trace back where the max pooling layer's features come from. For each dimension, the pooling layer chooses the maximum value from the nodes that are pooled to it. Thus, we can count the fraction in which a node's features are gathered by pooling. Intuitively, if a node's features are more related to the task, the fraction tends to be larger, and vice versa.

Figure 6.7 illustrates several example processed by d-TBCNN in the task of sentiment analysis. They correspond to the following sentences.

(a) *A masterpiece four years in the making*
(b) *In the process they demonstrate that there's still a lot of life in Hong Kong cinema*
(c) *The stunning dreamlike visual will impress even those viewers who have little patience for Euro-film pretension*

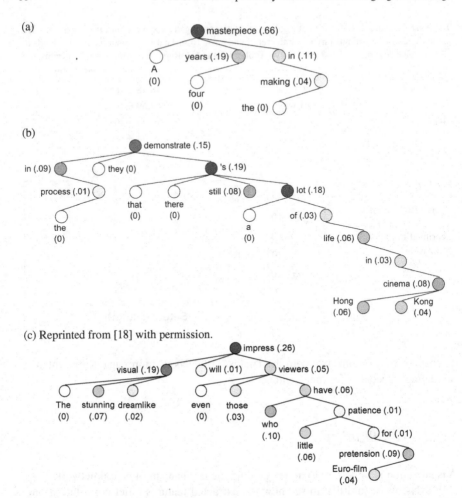

Fig. 6.7 Visualization of TBCNN features. These dependency trees correspond to: **a** "A masterpiece four years in the making"; **b**" In the process they demonstrate that there's still a lot of life in Hong Kong cinema"; and **c** "The stunning dreamlike visual will impress even those viewers who have little patience for Euro-film pretension." The number in the bracket indicates the percentage of convolution features being collected by the subsequent pooling layer. The ratio is also shown by color

Here, we applied one-way pooling because information tracing is more meaning-ful with one pooling slot.

As shown in the figure, tree-based convolution can effectively extract information relevant to the task of interest. We take Fig. 6.7 as an example. The two-layer windows corresponding to *"visual will impress viewers," "the stunning dreamlike visual,"* say, are discriminative to the sentence's sentiment. Hence, large fractions (0.24 and 0.19) of their features (after convolution) are gathered by pooling. On the other hand, words like *the*, *will*, *even* are known as stop words [6]. They are mostly noninformative for sentiment; hence, no (or minimal) features are gathered. Such results are consistent with human intuition.

We further observe that tree-based convolution does integrate information of different words in the window. For example, the word *stunning* appears in two windows: (a) the window "*stunning*" itself, and (b) the window of "*the stunning dreamlike visual*," with root node *visual*, and *stunning* acting as a child. We see that Window *b* is more relevant to the ultimate sentiment than Window *a*, with fractions 0.19 versus 0.07, even though the root *visual* itself is neutral in sentiment. In fact, Window *a* has a larger fraction than the sum of its children's (the windows of "*the*," "*stunning*," and "*dreamlike*").

6.4 Conclusion and Discussion

In this chapter, we propose the dependency tree-based convolutional neural network (d-TBCNN). In the dependency tree, each node corresponds to a word in the sentence. Therefore, we simply use word embeddings as nodes' vector representations. However, the nodes in dependency trees may have different numbers of child nodes. We extend the concept of convolution in d-TBCNN, as we index convolution weights by dependency types, instead of the positions within the convolution window. We also propose an "equal allocation" strategy for dynamic pooling.

We evaluated d-TBCNN in two sets of experiments: sentence classification and sentence matching. The first set has two tasks, sentiment analysis and question classification, and in both tasks, our model outperforms previous state-of-the-art results. In sentence matching, d-TBCNN achieves a better performance than sentence encoding-based approaches. Its performance is close to a word-to-word attention method, which has a higher complexity order than d-TBCNN. In addition, this chapter contains in-depth model analysis and visualization, so that we can have a better understanding of TBCNN.

We discuss the following issues.

Convolutional neural network versus recurrent neural network. Convolutional neural networks (CNNs) and recurrent neural networks (RNNs) are two widely applied standard neural models in NLP. It is mostly a consensus that RNNs are better than CNNs in most text generation applications, e.g., machine translation, abstractive summarization, and human–computer dialog systems. Nevertheless, CNN might be better than RNN in some scenarios, especially for discriminative classification without many data or when the sequence is too long. In our work, we incorporate the tree-structured prior into traditional CNNs and further improve the performance.

Dependency parsing versus constituency parsing. Intuitively, both c-TBCNN and d-TBCNN have their own advantage. c-TBCNN can better aggregate global information, as we have an underlying recursive neural network that learns non-leaf nodes' representation; d-TBCNN is more efficient in structure representation, as dependency trees are more compact. In our experiments, c-TBCNN and d-TBCNN yield similar results, d-TBCNN slightly outperforming c-TBCNN.

References

1. Bos, J., Markert, K.: Combining shallow and deep NLP methods for recognizing textual entailment. In: Proceedings of the First PASCAL Challenges Workshop on Recognising Textual Entailment, pp. 65–68 (2005)
2. Bowman, S.R., Angeli, G., Potts, C., Manning, C.D.: A large annotated corpus for learning natural language inference. In: Proceedings of the 2015 Conference on Empirical Methods in Natural Language Processing, pp. 632–642 (2015)
3. Braud, C., Denis, P.: Comparing word representations for implicit discourse relation classification. In: Proceedings of the 2015 Conference on Empirical Methods in Natural Language Processing, pp. 2201–2211 (2015)
4. Bromley, J., Guyon, I., LeCun, Y., Säckinger, E., Shah, R.: Signature verification using a "ssiamese" time delay neural network. In: Advances in Neural Information Processing Systems, pp. 737–744 (1994)
5. Collobert, R., Weston, J.: A unified architecture for natural language processing: deep neural networks with multitask learning. In: Proceedings of the 25th International Conference on Machine Learning, pp. 160–167 (2008)
6. Fox, C.: A stop list for general text. In: ACM SIGIR Forum, pp. 19–21 (1989)
7. Harabagiu, S., Hickl, A.: Methods for using textual entailment in open-domain question answering. In: Proceedings of the 21st International Conference on Computational Linguistics and the 44th Annual Meeting of the Association for Computational Linguistics, pp. 905–912 (2006)
8. Harabagiu, S., Hickl, A., Lacatusu, F.: Negation, contrast and contradiction in text processing. In: Proceedings of AAAI Conference on Artificial Intelligence, pp. 755–762 (2006)
9. He, H., Gimpel, K., Lin, J.: Multi-perspective sentence similarity modeling with convolutional neural networks. In: Proceedings of the 2015 Conference on Empirical Methods in Natural Language Processing, pp. 17–21 (2015)
10. Hu, B., Lu, Z., Li, H., Chen, Q.: Convolutional neural network architectures for matching natural language sentences. In: Advances in Neural Information Processing Systems, pp. 2042–2050 (2014)
11. Irsoy, O., Cardie, C.: Deep recursive neural networks for compositionality in language. In: Advances in Neural Information Processing Systems, pp. 2096–2104 (2014)
12. Jurafsky, D., Martin, J.: Speech and Language Processing. Pearson Education (2000)
13. Kim, Y.: Convolutional neural networks for sentence classification. In: Proceedings of the 2014 Conference on Empirical Methods in Natural Language Processing, pp. 1746–1751 (2014)
14. Lacatusu, F., Hickl, A., Roberts, K., Shi, Y., Bensley, J., Rink, B., Wang, P., Taylor, L.: LCCs GISTexter at DUC 2006: Multi-strategy multi-document summarization. In: Proceedings of DUC 2006 (2006)
15. MacCartney, B., Grenager, T., de Marneffe, M.C., Cer, D., Manning, C.D.: Learning to recognize features of valid textual entailments. In: Proceedings of the Human Language Technology Conference of the NAACL, pp. 41–48 (2006)
16. Mikolov, T., Yih, W.t., Zweig, G.: Linguistic regularities in continuous space word representations. In: Proceedings of the 2013 Conference of the North American Chapter of the Association for Computational Linguistics: Human Language Technologies, pp. 746–751 (2013)
17. Mou, L., Peng, H., Li, G., Xu, Y., Zhang, L., Jin, Z.: Discriminative neural sentence modeling by tree-based convolution. In: Proceedings of the 2015 Conference on Empirical Methods in Natural Language Processing, pp. 2315–2325 (2015)
18. Mou, L., Men, R., Li, G., Xu, Y., Zhang, L., Yan, R., Jin, Z.: Natural language inference by tree-based convolution and heuristic matching. In: Proceedings of the 54th Annual Meeting of the Association for Computational Linguistics, vol. 2, pp. 130–136 (2016)

19. Rocktäschel, T., Grefenstette, E., Hermann, K.M., Kočiský, T., Blunsom, P.: Reasoning about entailment with neural attention. In: Proceedings of the International Conference on Learning Representations (2016)
20. Silva, J., Cohéur, L., Mendes, A., Wichert, A.: From symbolic to sub-symbolic information in question classification. Artif. Intell. Rev. **35**(2), 137–154 (2011)
21. Socher, R., Pennington, J., Huang, E., Ng, A., Manning, C.: Semi-supervised recursive autoencoders for predicting sentiment distributions. In: Proceedings of the Conference on Empirical Methods in Natural Language Processing, pp. 151–161 (2011)
22. Tai, K., Socher, R., Manning, D.: Improved semantic representations from tree-structured long short-term memory networks. In: Proceedings of the 53rd Annual Meeting of the Association for Computational Linguistics, pp. 1556–1566 (2015)
23. Vendrov, I., Kiros, R., Fidler, S., Urtasun, R.: Order-embeddings of images and language. In: Proceedings of International Conference on Learning Representations (2016)
24. Yan, R., Song, Y., Wu, H.: Learning to respond with deep neural networks for retrieval-based human-computer conversation system. In: Proceedings of the 39th International ACM SIGIR Conference on Research and Development in Information Retrieval, pp. 55–64 (2016)

Chapter 7
Conclusion and Future Work

Abstract As the last chapter of this book, we will have conclusion in Sect. 7.1. Then we point of several future directions in Sect. 7.2, including graph-based neural networks, deep learning-based program analysis, and neural parsing.

Keywords Tree-based convolution · Structured data
Deep learning-based program analysis · Neural parsing

7.1 Conclusion

In this book, we propose the tree-based convolutional neural network (TBCNN). We analyze the advantage and disadvantage of traditional convolutional neural networks and recursive neural networks. Convolutional neural networks have short propagation paths, thus being able to learn features effectively. Recursive neural networks can make explicit use of sentences' internal structures (e.g., parse trees); however, it may suffer from the problem of gradient exploding or vanishing.

This book makes effort to combine the merits of both convolutional neural networks and recursive neural networks, and proposes the TBCNN model. Its core idea is to design a sub-tree sliding window, applied to different regions of the tree. The features extracted by tree-based convolution are aggregated by a dynamic pooling layer for further processing.

Table 7.1 compares TBCNN with existing models. The comparison involves two dimensions: (1) the way of information propagation, including iterative information aggregation and sliding window for information extraction; and (2) the structure of data, including "linear" or "flat" data, and tree-structured input. Located in the table are the conventional convolutional, recurrent, and recursive neural networks, as well as the TBCNN proposed in this book. In particular, TBCNN uses sliding windows to extract tree structural features.

We apply TBCNN to two domains: program analysis and natural language processing, and specifically, we consider abstract syntax trees, constituency trees, and dependency trees. When designing TBCNN variants for different types of trees,

© The Author(s) 2018
L. Mou and Z. Jin, *Tree-Based Convolutional Neural Networks*,
SpringerBriefs in Computer Science, https://doi.org/10.1007/978-981-13-1870-2_7

Table 7.1 Position of TBCNN among related models. We compare TBCNN with recurrent/recursive neural networks and convolutional neural networks in two aspects: (1) the way of information propagation, and (2) the sentence structure that the network uses

Structure	Way of information propagation	
	Iterative	Sliding
Flat	Recurrent	Convolution
Tree	Recursive	Tree-base convolution

we mainly consider three technical difficulties, namely, the vector representation of nodes, the weight of convolution, and the dynamic pooling.

In our book, we evaluated the effectiveness of TBCNN in various tasks, including program functionality classification, program snippet pattern recognition, natural language sentiment analysis, question classification, and natural language inference (NLI). TBCNN outperforms previous state-of-the-art models in all tasks except NLI. For NLI, TBCNN also outperforms existing sentence encoding-based approaches, and achieves similar accuracy to a word-by-word attention model, whose matching complexity is higher than TBCNN in order.

7.2 Future Work

Despite the flexible variants and wide applications of TBCNN discussed in this book, we would also like to point out several future directions of research related to our work.

Deep learning for graph structured data. Using neural networks to analyze structured data has been increasing popular recently. Our book deals with tree structures, and other interesting data structures—generally speaking, graphs—include molecules [1] and social networks [2].

The idea of tree-based convolution can be directly extended to graph convolution [3, 4]. Moreover, graphs have a more solid mathematical foundation, allowing convolution in the spectral domain [5].

Other learning algorithms traditionally working with "flat" structures can also be extended to trees or graphs in a similar way. For example, DeepWalk redefines neighboring by graph connections and extends sequential word embedding learning to the graph structure [2]. Choi et al. [6] extend the attention mechanism to graph structures.

Deep learning for program analysis. In Chap. 4, we apply deep learning to program analysis. As a starting point, we mainly consider program classification based on abstract syntax trees (ASTs). Followup work addresses binary code [7], program identifiers [8], application programming interfaces (APIs) [9], and other program components.

Besides classification problems considered in this book, other studies explore various exciting applications, including program summarization [10], bug locating [11], and code search [12]. We have also explored neural program generation [13] and SQL-like neural program execution [14] in our subsequent work.

Neural Parsing. The last direction we would like to mention is neural parsing. In this book, our networks make use of parsing results by building a tree convolutional layer along the parse structures. In fact, neural networks can be trained to predict parsing structures. Such problem would be easy if groundtruth parsing labels are available for the training data [15, 16]. A more challenging and interesting setting of neural parsing is that groundtruth parsing labels are unknown even in the training set. In this case, the network is trained by some downstream tasks (e.g., language modeling, sentence classification), but people would anticipate that parsing structures can emerge by proper neural designs [17, 18].

References

1. Xu, Y., Dai, Z., Chen, F., Gao, S., Pei, J., Lai, L.: Deep learning for drug-induced liver injury. J. Chem. Inf. Model. **55**(10), 2085–2093 (2015)
2. Perozzi, B., Al-Rfou, R., Skiena, S.: DeepWalk: online learning of social representations. In: Proceedings of the 20th ACM SIGKDD International Conference on Knowledge Discovery and Data Mining, pp. 701–710 (2014)
3. Duvenaud, D.K., Maclaurin, D., Iparraguirre, J., Bombarell, R., Hirzel, T., Aspuru-Guzik, A., Adams, R.P.: Convolutional networks on graphs for learning molecular fingerprints. In: Advances in Neural Information Processing Systems, pp. 2224–2232 (2015)
4. Marcheggiani, D., Titov, I.: Encoding sentences with graph convolutional networks for semantic role labeling. In: Proceedings of the 2017 Conference on Empirical Methods in Natural Language Processing, pp. 1506–1515 (2017)
5. Kipf, T.N., Welling, M.: Semi-supervised classification with graph convolutional networks. In: Proceedings of the International Conference on Learning Representations (2016)
6. Choi, E., Bahadori, M.T., Song, L., Stewart, W.F., Sun, J.: GRAM: graph-based attention model for healthcare representation learning. In: Proceedings of the 23rd ACM SIGKDD International Conference on Knowledge Discovery and Data Mining, pp. 787–795 (2017)
7. Shin, E.C.R., Song, D., Moazzezi, R.: Recognizing functions in binaries with neural networks. In: Proceedings of the 24th USENIX Security Symposium, pp. 611–626 (2015)
8. Allamanis, M., Barr, E.T., Bird, C., Sutton, C.: Suggesting accurate method and class names. In: Proceedings of the 2015 10th Joint Meeting on Foundations of Software Engineering, pp. 38–49 (2015)
9. Gu, X., Zhang, H., Zhang, D., Kim, S.: Deep API learning. In: Proceedings of the 2016 24th ACM SIGSOFT International Symposium on Foundations of Software Engineering, pp. 631–642 (2016)
10. Allamanis, M., Peng, H., Sutton, C.: A convolutional attention network for extreme summarization of source code. In: Proceedings of the International Conference on Machine Learning, pp. 2091–2100 (2016)
11. Huo, X., Li, M.: Enhancing the unified features to locate buggy files by exploiting the sequential nature of source code. In: Proceedings of the 26th International Joint Conference on Artificial Intelligence, pp. 1909–1915 (2017)
12. Gu, X., Zhang, H., Kim, S.: Deep code search. In: Proceedings of the 40th International Conference on Software Engineering, pp. 933–944 (2018)

13. Mou, L., Men, R., Li, G., Zhang, L., Jin, Z.: On end-to-end program generation from user intention by deep neural networks (2015). arXiv preprint arXiv:1510.07211
14. Mou, L., Lu, Z., Li, H., Jin, Z.: Coupling distributed and symbolic execution for natural language queries. In: Proceedings of the 34th International Conference on Machine Learning, pp. 2518–2526 (2017)
15. Bowman, S.R., Gauthier, J., Rastogi, A., Gupta, R., Manning, C.D., Potts, C.: A fast unified model for parsing and sentence understanding. In: Proceedings of the 54th Annual Meeting of the Association for Computational Linguistics, pp. 1466–1477 (2016)
16. Zhou, H., Zhang, Y., Huang, S., Chen, J.: A neural probabilistic structured-prediction model for transition-based dependency parsing. In: Proceedings of the 53rd Annual Meeting of the Association for Computational Linguistics and the 7th International Joint Conference on Natural Language Processing, pp. 1213–1222 (2015)
17. Shen, Y., Lin, Z., Huang, C.W., Courville, A.: Neural language modeling by jointly learning syntax and lexicon. In: Proceedings of the International Conference on Learning Representations (2017)
18. Williams, A., Drozdov, A., Bowman, S.R.: Do latent tree learning models identify meaningful structure in sentences? Trans. Assoc. Comput. Linguist. **6**, 253–267 (2018)

Index

© The Author(s) 2018
L. Mou and Z. Jin, *Tree-Based Convolutional Neural Networks*,
SpringerBriefs in Computer Science, https://doi.org/10.1007/978-981-13-1870-2

Printed in the United States
By Bookmasters